FULLER, John Grant, 1913-

We almost lost Detroit [by] John G. Fuller. N.Y., Reader's Digest Pr.; distrib. by T. Y. Crowell [c]1975.
272p.
Includes bibliography.

1. Enrico Fermi Atomic Power Plant, Michigan
2. Atomic power-plant - Accidents I. tc.

PMS0590-80 CIP

WE ALMOST LOST DETROIT

Other Books by John G. Fuller

THE GENTLEMEN CONSPIRATORS
THE MONEY CHANGERS
INCIDENT AT EXETER
INTERRUPTED JOURNEY
THE DAY OF ST. ANTHONY'S FIRE
THE GREAT SOUL TRIAL
200,000,000 GUINEA PIGS
FEVER!
ARIGO: SURGEON OF THE RUSTY KNIFE

WE ALMOST LOST DETROIT

JOHN G. FULLER

READER'S DIGEST PRESS
Distributed by Thomas Y. Crowell Company, New York, 1975

 Published simultaneously in Canada by Fitzhenry & Whiteside Limited, Toronto.

Manufactured in the United States of America

Library of Congress Cataloging in Publication Data

Fuller, John Grant, 1913–
We almost lost Detroit.

Bibliography: p.
Includes index.
1. Enrico Fermi Atomic Power Plant, Michigan. 2. Atomic power-plants—Accidents. I. Title.
TK1344.M5F84 1975 621.48′3 75-17870
ISBN 0-88349-070-6

1 2 3 4 5 6 7 8 9 10

FOREWORD

Nuclear power is an unforgiving technology. It allows no room for error. Perfection must be achieved if accidents that affect the general public are to be prevented.

WE ALMOST LOST DETROIT is a valuable addition to the literature of the debate swirling around nuclear power. The book is well researched; it provides fascinating reading for anyone who is interested in the history of the nuclear program. The author provides an in depth review of the development of the Fermi liquid metal fast breeder reactor, using the Fermi accident as a vehicle to discuss the significance of the entire nuclear power industry, both in this country and abroad. In the course of this absorbing story, the causes and consequences of a number of nuclear accidents are described in a manner that is easily understood. For this is not a "technical" book meant only for the scientific community. It is written in a free flowing style that can be easily followed by the layman.

This book deals primarily with the accident at the Fermi No. 1 liquid metal fast breeder reactor and the safety problems the accident posed. One result was to set back the liquid metal fast breeder program by many years. However, the government is still promoting new designs of these reactors for use in the next decade, and the problems exposed in this book will not disappear. Meanwhile, the light-water reactors that are currently being used in this country have safety problems of their own, as this book demonstrates. For example, a light-water reactor cannot undergo a low-order nuclear explosion as can occur in a breeder reactor. But a loss-of-coolant accident in a light-water reactor could lead to a core meltdown. Thus, while the accident sequences are different for the two types of reactors, the end result, the possible release of radioactive material into the environment—is the same.

The developers of the Fermi breeder reactor were very sincere, diligent, and highly qualified individuals to whom the safety of the reactor was paramount. Extreme care was taken to insure against the possibility of a serious accident occurring. The scientists involved were most confident that they had covered all possible problem areas. They had built safeguards on top of safeguards. Yet in spite of the precautions in the design and construction of the Fermi reactor, and in spite of the reassurances by the scientists that a serious accident could not happen, one did occur. The results far exceeded the expectations of anyone involved with the project. Fortunately, at the time of the accident, the reactor was operating at a very low power level or the consequences could have been much worse.

The Fermi accident and the others described in this book demonstrate the fact that no matter how much diligence is exercised in the design, construction, and operation of a nuclear reactor things can and do go wrong. Design errors occur, the unexpected happens, human error is a very real possibility.

The nuclear industry in this country today is experiencing many safety-related dilemmas. Several recent problems have raised questions about the mechanical integrity of the piping systems. Construction quality control has not been good. Emergency diesel generator units have failed frequently during tests. In one recent instance, two out of three such safety units failed during a test: one because of a faulty timer switch and the other because of water in the fuel tank. In one incident the high-pressure emergency-core-cooling system failed to operate when called upon because several diodes had burned out. In that same incident a four-foot-long piece of pipe that had been inadvertently left in the reactor during construction jammed in a valve and prevented it from closing. In another, and one of the most serious accidents to occur in a large commercial power plant, a fire in an electrical cable duct knocked out numerous electrical circuits, many of them redundant circuits. The emergency-core-cooling system was incapacitated. If it had been needed it would not have been available. These examples are listed merely as an indication of the type of problems that can arise.

It is interesting to note the recent shift in the nuclear industry's position about the possibility of nuclear power plant accidents. For many years, the industry vigorously defended the nuclear power program as being essentially risk-free. Nuclear power was claimed to be perfectly safe. It was said that no serious accidents would ever happen. Such a position was of course necessary to promote the acceptance of nuclear power by the general public. It has not been until just recently that the proponents of nuclear energy have admitted that accidents can and will happen, and the public should prepare itself for such eventualities.

What is forgotten is that the public had a right to know these risks years ago, when the initial decisions regarding the acceptability of nuclear power were being made. Only now are we learning that the public was deliberately misled and deceived by the former Atomic Energy Commission regarding the possibility of major nuclear reactor accidents and the potential consequences of such accidents.

The discussion in this book of the WASH-740 update report

of 1964 is very enlightening. The reader is provided an insight into the internal workings of the AEC by the quoting of letters and memos of the AEC officials and consultants who were involved in the study. What this book shows is that the estimated consequences of a major accident were so frightening the AEC and nuclear industry did not want them revealed to the public. It was the fear of an adverse public reaction that prompted the AEC to suppress the report, and this is a good example of the attitude that has prevailed in the nuclear power program. The public has been deliberately misled into believing that there are no problems with nuclear power, whereas there are very real problems. In addition to the issue of reactor safety, other problem areas include transportation, fuel processing, waste disposal, nuclear material diversion, and sabotage.

In more than seven years of working with the AEC's safety research program for light-water reactors, I had an excellent opportunity not only to become familiar with the AEC's research programs and safety analysis methods, but also to observe the basic underlying philosophy of the AEC. This attitude was primarily one of trying to prove that existing reactors were safe rather than one of independently assessing the adequacy of the safety systems. While many of the scientists working on the safety research were conscientious and tried to point out valid problems regarding reactor safety, their questions were largely ignored. The decisions regarding safety research programs were made by the AEC in Washington, not by the scientists in the laboratories. Worse, many of the managers in private industry that ran the laboratories for the AEC were more interested in keeping their contracts than they were in doing the research as it should have been done. The managers' philosophy was that the AEC was always right.

I left my job with Aerojet Nuclear Company, the AEC's major safety contractor at the Idaho National Engineering Laboratory, because of a growing frustration with the safety program. I became particularly concerned about the way in which the AEC had continually misled the public about the safety of nuclear reactors. Only favorable results regarding the safety research were reported. I knew well the large number of uncertainties and problems that were not freely publicized; only a continuing pressure from citizen groups has made these uncertainties known to the general public. And I am concerned that the safety systems on the reactors operating in this country have not been tested, and the adequacy of these systems has yet to be proven.

The AEC, as a government agency, had an obligation to serve the public in an unbiased manner. It did not. The new Nuclear Regulatory Commission has been formed specifically to regulate, and not promote, the nuclear industry. But after the first several months of operation, there does not appear to be a truly unbiased view prevalent in the NRC.

The public has a right to know all the pertinent information regarding not only the safety involved in nuclear reactors, but all the other related problems. Only then can a rational decision regarding the acceptability of nuclear power be made.

Carl J. Hocevar
Union of Concerned Scientists
Cambridge, Mass.

WE ALMOST LOST DETROIT

ONE

The phone call came in sometime in the mid-afternoon of Wednesday, October 5, 1966. The exact time is not recorded, because it was never entered officially on the log of the sheriff of Monroe County, Michigan. Sheriff Charles Harrington, known as Bud, a lanky man with a lean, craggy face, received it. An unidentified voice on the other end of the line spoke sharply and briefly, saying it was Detroit Edison calling—the major utility company in southeastern Michigan. There was something wrong at the new Enrico Fermi Atomic Power Plant, which Detroit Edison operated at Lagoona Beach—just a handful of miles away from the town of Monroe. The cause of the problem was uncertain, but the caller said that the situation should not be publicized, that no public alert should be given. More information would follow.

Sheriff Harrington hung up the phone in his tiny office, over-crowded by just one desk and his radio-communications equipment. He went directly into the next door office of the chief of police of the town of Monroe. Both agreed they would not enter the information on the blotter, and would keep it to themselves. Both men knew that a nuclear power plant contains within it more potential radioactive fallout than dozens of Hiroshima-type A-bombs. The Fermi plant was no exception: It was new, it was untried, it was being tested. And both knew that the ultimate action in case of a major atomic plant accident was evacuation. Yet, if the public was given any hint of the problem there could be mass panic. The two officers decided to wait it out.

At about the same time, some one hundred miles away, Captain Buchanan of the Michigan State Police in Lansing was alerted by a similar phone call, again from a Detroit Edison representative. The state police are responsible for the civil defense of the state. But neither Captain Buchanan nor Sheriff Harrington—nor even the nuclear experts—knew all the possible consequences of a civilian atomic power plant accident. In 1966, there were only four commercial nuclear power plants in the entire country. There was no experience to act on.

The only certainty was that the escape of radioactive poisons due to a major reactor accident could be catastrophic. The Atomic Energy Commission's (AEC's) own figures were already available in a study known as WASH-740. In the worst case, for a relatively small reactor, the AEC estimated 3,400 people could be killed, 43,000 could be injured or stricken with radiation poisoning, $7 billion in property damage could occur, and an area the size of Pennsylvania could be contaminated. These figures, however, were already outmoded. A new study that nestled in the "Official Use Only" files of the AEC went far beyond these estimates.

In any case, the only action specified by the AEC in the event of such an accident was evacuation of the population from the surrounding area. This was not a source of comfort to Sheriff Harrington, Captain Buchanan, or anyone else responsible for coping with the situation.

Hardly anyone in Monroe County was giving any thought that day to the spanking new Enrico Fermi Atomic Power Plant at Lagoona Beach on the shore of Lake Erie. The citizenry was apathetic about it, except for the jobs it provided and the new tax revenues it brought in. Everyone living in the clustered towns of Monroe, Frenchtown, and Newport, surrounding Lagoona Beach, had grown used to the old cumbersome Detroit Edison coal-fired plant near the lake. They had come to ignore the almost grotesque procession of giant transmission towers that seemed to promenade across the flatlands alongside the Dixie Highway, like skeletons of enormous scarecrows dragging festoons of high-tension wires.

The tang and wild beauty had long since gone out of the marshes and wetlands near Lagoona Beach, once smothered with lotus and wild rice. The wastes from paper mills and steel companies joined with other sewage in the lake; *Unsafe-for-Swimming* signs were everywhere.

Monroe and its satellite towns were undergoing economic doldrums. Only an occasional festive muskrat dinner brightened things up. Legend has it that when Monroe's favorite son, General Custer, had left town on his way to the disaster at the Little Big Horn, he had admonished the citizens not to do anything until he got back. Some of the more dour residents today claim that these instructions have been followed to the letter. About the only occurrence of public note that October 5 was that Hubert Humphrey, on a mission to dedicate the Monroe Public Library, had arrived fifty-one minutes late at the Custer Municipal Airport. Aside from that, the community was listless.

An early morning frost had covered the corn and soybean crops throughout the county, but the county extension agent didn't seem too worried about it, though he was concerned about the tomatoes and peppers. Dan Warner of Ida Orchards wasn't at all concerned. "Frost is good for the apples," he said. "It helps them ripen."

About a half hour's drive north from Monroe was Detroit. The motor city went about its business putting cars, trucks, and tractors together, but not all was happy there. More than 40,000 Chrysler workers were about to be laid off; holdups and burglaries were rising at an alarming rate, up 7,000 cases from the previous year. Round steak was selling at 89¢ a pound at Kroger's, with sirloin going at a dime higher. Movie goers filed in to see *Dr. Zhivago* at the United Artists Theater, and *The Sound of Music* at the Madison.

Across the Detroit River, directly south and east, Windsor, Ontario, went about its job of turning out steel, machinery, and chemicals, as bored Canadian customs officials waved an endless procession of cars through the tunnel to and from Detroit.

Farther out on the western rim of the circle, Dearborn, Pontiac, Willow Run, Ypsilanti, and other satellite towns of Detroit were equally placid, dwelling mainly on ordinary problems of everyday routine. In these and other suburbs, housewives flooded the shopping centers, and visitors moved peacefully in and out of the Henry Ford Museum in Greenfield Village. Few in any of these communities knew much about the Enrico Fermi Atomic Plant, except that it was supposed to promise more electric power through some nuclear magic. And, of course, it also provided jobs.

At 5739 Nelson Drive in Monroe, barrel-chested Frank Kuron, an ironworker at the plant, was preparing to drive his wife to the supermarket. He also wanted to pick up a fifth of Canadian Club and a six-pack of Miller Hi-Life. Kuron, with a rough, raw voice and an astoundingly good vocabulary, has a capacity to stay cold sober when downing his after-work boilermakers (a shot of Canadian, a swallow of beer). He is chunky and solid, with a pleasant, gnarled face. He likes to call himself The Polack, likes to follow current events, likes to read books on the economic and political scene, likes to salt away his money.

Kuron had been on the first gang that unloaded the red iron, those bright-orange structural steel girders that created the framework of the Fermi atomic plant buildings. The girders had arrived by truck and barge some ten years earlier. And from the time he had first picked up his pneumatic impact wrench (he called it a yo-yo) to bolt the cold-bent plates, beams, and the triangular gussets into place, Kuron swore by the craftsmanship that went into the building. Most of his working days over the past ten years had been spent on the Fermi job. He had nothing but praise for the way it was put together. He even had praise and respect for the "blue hats"—the management engineers whose hard hats were colored blue to distinguish them from the orange hats of the subcontractors. There was a sharp distinction between the blue hats and the white hats with their various stripes of colored tapes that designated those of a laborer (a yellow "X"), ironworker (a pink stripe), mason (gray), carpenter (blue), and others.

For a decade, Kuron had watched the Fermi plant grow from its red-iron skeleton into a new, gleaming white cluster of buildings on the edge of Lake Erie, less than two miles from his home. In between the low, squat office section and the gaunt, looming cracker box that housed the turbogenerators was the shiny white dome of the nuclear reactor building. Underneath the dome was the nuclear reactor itself, the atomic furnace that was just now, in October of 1966, at the point of heralding a hopeful new era of the peaceful atom for the Detroit area.

In the course of assigning the responsibilities for evacuating communities in case of radiation contamination, planners could not foresee practical problems that would thwart their efforts.

For the Michigan State Police, who bore the responsibility for the whole state, the task of evacuating Detroit would be flatly impossible because the automobile city had put all its faith in motor transport. It had never built a subway or elevated transit system. During commuting hours, traffic was impossible. The streets were always jammed. The freeways were worse in spite of an imposing web of them, including the Edsel Ford, the Detroit Industrial, the Southfield, and the Fisher Freeways—the latter being Interstate 75, the four-lane artery that led south to Monroe and Toledo, Ohio. Coordination with the Canadian Civil Defense in Windsor would be futile. The tunnels and bridges would be packed to capacity.

Even in rural Monroe County, there would be 200,000 people in the area for Sheriff Harrington to worry about, more than 20,000 of them in the town of Monroe itself. The winds of Lake Erie could be fickle, often embracing every point of the compass over a period of days. If those shifting winds carried the colorless, odorless, tasteless radioactivity, evacuation would be more difficult.

Most vulnerable were the houses scattered in a two-mile radius around the site boundary. There were about 120 of these around Port Sunlight; another 100 or so inland from Swan Creek; 75 along Pointe Aux Peaux Road, and 25 between the Fermi plant and Dixie Highway. Another 500 houses—mainly cottages—sprawled along the narrow roads of Stony Point where Frank Kuron lived. Once a resort area, it was now converted to year-round living. It lay behind a massive rocky breakwater that kept out the surly chop of Lake Erie. If it came to evacuation, the residents of Stony Point had little choice. They would have to make their way directly toward the Pointe Aux Peaux Road in the direction of the Fermi reactor, and move precariously along the site boundary to whatever haven they could find from the ghostly plume of radiation that might spring from the reactor building if its containment shell were broken.

Back in his home on Stony Point later that afternoon, Frank Kuron relaxed on his overstuffed sofa, oblivious to what was taking place less than two miles away from his home. Nor was he alone in his ignorance. The problem was that no one, not even the engineers and operators in the Fermi control room, knew exactly what was happening—aside from the fact that at 3:09 that afternoon the radiation alarms had sounded, a Class I alert had been issued, and the reactor building had been automatically shut off from the rest of the complex. What was to follow was anybody's guess. But whatever it might be, it had been in the making for more than a decade.

Some fifteen years before the radiation alarms went off, the groundwork for the Fermi plant had been laid in Detroit. In December of 1951 Walker L. Cisler had just been named president of the Detroit Edison Company, one of the most prestigious and largest utilities in the Midwest.

Cisler was both an idealist and a pragmatist, a dreamer with visceral strength and an abrasive magnetism who seemed to get things done. As an idealist, he extolled the virtues of the human mind, claiming that it was the greatest motivator on earth aside from God. He was a Cornell engineering graduate of the class of 1922, and he had a handsome, impassive face, with classic, chiseled features framed by cropped white hair. His social consciousness was reflected in his overseas work as a consultant on electric power for the Marshall Plan, and the Agency for International Development, plus a half a dozen other government agencies. He engaged in scores of civic activities in Detroit. His dedicated service to the public, however, was tempered by an obsessional passion to ram through projects of his liking at almost any cost. Some of his associates felt that his positivism was often blind and overpowering. But it got the desired results.

Besides being named president of Detroit Edison, Cisler had

just delivered a feasibility study to the Atomic Energy Commission. It concluded that a nuclear power plant would not only be practical for private enterprise, but could lead to a whole new concept of meeting energy needs for an uncertain future.

Cisler had been working on the idea for more than four years, from the time he had been appointed by the AEC as a member of an advisory group. He had received an AEC blessing to go ahead with a joint plan between Detroit Edison and Dow Chemical to examine exactly how nuclear power could be tamed.

Back in 1951, experts believed the future of nuclear power could be almost miraculous: A reactor could be built that would not only create heat to power electrical generators, but would produce more fissionable fuel than it used. It would be called the breeder reactor, which meant just what it said.

The technique involved packing a blanket of raw, sluggish Uranium-238 around a reactor core. This would capture the excess neutrons that leaked out of the more refined Uranium-235 in the center as the fuel went through the fission or splitting process. While the heat was removed from the core to make steam for electric power, the leaking neutrons would smash into the blanket and change the Uranium-238 into Plutonium-239, an able and superelegant fuel, but one that was not without its drawbacks. As a new, man-made element, plutonium was not only the source of the explosion of the nuclear bomb, but was also perhaps the most deadly poison in the world. Once created, it would take 480,000 years before its radioactivity would decay to an innocuous radiation level.

But the dreams of Cisler and other engineers and scientists across the country, working to bring about a brighter world in the closing days of 1951, were not bent on destruction. These were men of hope, vision, enterprise, brilliance, and determination. They were bent on taming this poison if they could, harnessing it for the benefit of mankind. But like everyone else, they were also fallible.

The Atomic Energy Commission's blessing for the Edison-Dow plan to proceed to a "design phase" came through on December 19, 1951. It came only one day before the historic event that was to be recorded on the broad, rocky flats of Idaho, some 2,000 miles to the west of Detroit.

That event took place in one of a lonely cluster of buildings, among the first erected in the AEC's National Reactor Testing Station, not far from Arco, and only forty miles away from the town of Idaho Falls. Near the desolate national monument known as the Craters of the Moon and sandwiched between the Lost River range and Big Southern Butte, the site was later to be designated as a Registered National Historic Landmark.

Here, on December 20, 1951, sixteen tense men, under the direction of the nuclear pioneer Walter Zinn, waited for the results of their work on the nuclear power unit known as Experimental Breeder Reactor No. 1, to become known as EBR-I. While they watched, the control rods were gingerly withdrawn from the nuclear core by a mechanical overhead linkage, releasing the neutron population of its Uranium-235 fuel rods. The heat from the rods created steam, and suddenly four ordinary 200-watt light bulbs began to glow. Walter Zinn was the first to scrawl his name on the wall, followed by the fifteen other scientists and technicians who signed in under the improvised legend: *Electricity Was First Generated Here From Atomic Energy on December 20, 1951.*

The modest success of EBR-I was a harbinger of hope for Walker Cisler and his project. But there were marked differences to consider. EBR-I was a sputtering firecracker compared to the howitzer that a commercial reactor serving a public utility would have to be. EBR-I supplied enough electric light for one small building. Cisler's new plant would have to produce some 200,000 kilowatts of power to make any sense at all, even as a demonstration model. And, at best, it would be barely enough to handle 60,000 homes.

There were other considerations, too. EBR-I sat in a sparsely populated area. If anything went wrong, its small size, minute power, and remoteness would cause a minimum of damage to property and people. Cisler's new plant would be sitting in the middle of a teeming megalopolis.

To Cisler, there was another factor in the historic event at EBR-I. It was a government-owned reactor, wholly designed and built by the AEC. Private industry was playing only a supply role in its operation. Cisler believed devoutly in private industry, with its incentives and its initiative. Originally, he had dedicated his consulting work with the AEC "to speed industrial activities in the atomic energy field, including the development of atomic-electric power plants." He was aware, as was everyone else, that the wartime function of nuclear weapons could be handled in no other way but by an enormous government program. Power plants, he felt, were different.

Even with the coming of peace, there were weapon priorities that had to be faced in the arms race with Russia; there was strictly classified material to deal with, secrets to guard. Cisler was wary of government monopoly or encroachment on energy production. However, it seemed to create in him the fire of Sir Galahad in search of the Holy Grail. Armed with the authorization from the AEC, he set out on his mission to bring together a viable group of utilities that would join with him in mastering the intricacies of the new art. This would also bring enough private capital into the project, and demonstrate to both the AEC and the Joint Committee on Atomic Energy of Congress that the private sector meant business, and could do a better job than bureaucracy in developing the promise of nuclear power. Cisler did his utmost to convince both manufacturers and utilities that they *must* join in the battle, even if the ultimate profits might be a long way off in the future.

It took ten months, but his power and persuasion paid off. By October of 1952, he had created a Nuclear Power Development Department at Detroit Edison, and had managed to bring fifteen other utilities into the Edison-Dow breeder reactor project. They included some of the best blue-ribbon utilities in the country: from Consolidated Edison of New York to New England Power to Philadelphia Electric. By October 19, 1952, the AEC had approved the growing membership in the project—a unique cooperative venture that no individual utility could handle alone, in either money or expertise.

There was no question that both a heavy expenditure of money and solid expertise were needed if private companies were to catch up with the government's enormous nuclear establishment that had grown out of wartime necessity.

In December of 1952, however, the entire nuclear fraternity received a jolt. It was enough to make many sober scientific minds wonder whether this new technology that seemed to hold so much promise for the future of man was worthy of pursuit.

The Canadian government's atomic research center, some two hundred miles northwest of Ottawa, sits near the lonely outpost village of Chalk River, Ontario. Life there, in the close-by bedroom village of Deep River, can be pleasant for the scientists and other workers, even if remote from the mainstream of metropolitan living. There are the Staff Hotels for visitors; there are also schools, clubs, churches, and one movie house. On a hotel bulletin board, the notices posted give a clue to the life there: *Curling Is Fun—Try It This Winter. Trailer for Sale—Suitable for Snowmobile. Winter Carnival—February 7 to 9.* A sharp word to those who posted notices dominates the board: *All Notices Must Be*

Dated and Will Be Removed After 14 days. To which a whimsical scientist had scrawled as a postscript: *Date Is Missing from This Notice.*

As World War II drew to a close, Chalk River was selected from several alternate sites for the Canadian nuclear energy project. The reasons for its choice were specifically spelled out by the Canadian government. The site had to be isolated in case of an explosion, or emission of radioactive dust into the atmosphere. It had to be some distance from a town or a village. It had to have an ample supply of water to cool the reactor.

Chalk River easily met the requirements and the NRX experimental reactor was built on December 12, 1952. While Walker Cisler continued to cajole, persuade, beg, and convince his colleagues in the utility industry to support his plans for the future, the NRX reactor at Chalk River was going through a normal series of tests. They were experiments conducted at low power to compare the difference in reactivity (the ability to sustain a chain reaction) between fuel rods that had been bombarded by a long period of radiation and fresh fuel rods that had just been installed in the reactor.

The Canadian NRX reactor was also a midget compared to the one Cisler's group had in mind. It was what is known as a heavy-water reactor. Heavy water is a water molecule containing a hydrogen isotope with a mass number greater than 1. It is present in natural water about 1 part in 6,500 and is extremely expensive to isolate but a highly effective moderator. The heavy-water reactors are generally considered safer than the light-water reactors (which use plain, natural water as a coolant) on the drawing boards in the United States. Canada preferred the former not only for safety reasons, but because raw Uranium-238, which it had in plentiful supply, could be used. At the time, heavy-water reactors were considered safer than the breeder reactors, which were still in the early experimental stage. So safe, in fact, was the design of the NRX, that it had some nine hundred devices for shutting it down in an emergency and only one for starting it up.

In any kind of reactor, the fuel (most often uranium) is encased in metal sheaths that look very much like curtain rods. In some reactors, the rods are thick, and chunks of fuel, looking not unlike blackish Tootsie Rolls, are slipped into the rods. Some fuel rods are thin, with the metal sheath bonded to become part of them, depending on the type of the reactor. The rods are then bunched in bundles, called subassemblies. These are placed geometrically into a circular or octagonal container inside the reactor vessel, and packed like a round tin of cigarettes except for strategic spacing between the fuel elements. The fuel makes up the heart of the reactor—the core.

At strategic points among the fuel bundles, there are several long slender control rods. They are composed of cadmium or boron or graphite or some other material to stop the nuclear reaction. When they are plunged down inside the core, they drink up the fluctuating neutrons like a blotter and shut the reactor down. When they are lifted out of the core, the neutrons from the fuel begin sputtering (though invisibly and silently) like a massive pack of Fourth of July sparklers in a bucket. The neutrons then will crash into and split the atoms in the fuel bundles. The splitting not only gives off an enormous amount of heat, but sends two and a half more neutrons out of an atom's nucleus to repeat the process.

Whatever kind of coolant is used—light water, heavy water, or liquid sodium—it keeps the fuel from melting from its own heat. The coolant liquid bathes the hot fuel rods as it flows around them, carrying away their heat, which is then used to generate steam. The steam spins the turbines; the details of steam generation vary with reactor designs.

Being experimental, the NRX reactor at Chalk River was not designed for generating electric power. It was moderated by heavy water, which meant that it used this rare and expensive substance to slow down the neutrons so they would have a better chance of hitting more atoms, thus enabling it to use the natural Uranium-238 as fuel.

In the NRX control room on December 12, 1952, were a dozen or so men, including the project head, research physicists, a health physicist, operating superintendents, and reactor operators. At about three that afternoon, they were set to commence the routine experiments of the day. The atmosphere was relaxed, and because the experiments were being conducted at low power, there was little or nothing to worry about. Danger increases in proportion to the amount of power created. One small concern was that a special safety circuit was not in operation at the time, but since it was known, allowances would be made for this during the routine.

Just before the start of the experiment, an assistant operator in the basement below the reactor opened, by mistake, four valves that kept air pressure from raising the control rods after the initial start-up. If the rods were to rise the fuel in the reactor would immediately begin splitting atoms at a faster, unplanned rate, leading to one of the most feared circumstances of all—a nuclear runaway. From that point on it would be impossible to predict what would happen.

At the control panel, the supervisor of the operation was horrified when he saw red warning lights suddenly flash on the panel board. He grabbed the phone, yelled to the operator in the basement to stop what he was doing, then rushed down to the basement, leaving his assistant in charge of the panel.

In the basement, he was relieved to find that not all the valves were open. He closed the valves immediately, and was sure the control rods had returned to their proper position.

He checked the air pressure, and it was good. Up in the control room the red warning lights went off, indicating that the control rods were where they should be.

What he did not know, nor did anyone else at the time, was

that in some inexplicable way the rods had jammed, and had dropped down just far enough to turn off the warning lights, but not far enough to choke off the reactivity which was rising rapidly.

When the supervisor realized what was happening, he grabbed the basement phone to call the control room, intending to order his assistant to push buttons numbered 4 and 3 to stop the reactivity. Instead, he said: "Push number 4 and number 1."

Up at the control panel, in order to reach the two buttons, the assistant had to put the phone down. The moment he did so, the supervisor in the basement realized he had called out the wrong numbers. He yelled into the phone but no one heard him. The reactor began to run out of control—"above critical" in the parlance of the nuclear engineer.

It took only twenty seconds to realize this. Meanwhile, the power of the NRX reactor was doubling every two seconds. By that time, the reactor was on its way to a fuel meltdown. Four banks of control rods had been raised when the assistant had pushed button number 1. He immediately took the prescribed safety measure: He scrammed the reactor, tripping it so that all the control rods would slide safely back into place.

But, because of a lack of air pressure the control rods were not forced back into place. The galvanometer, which measures the electric current, indicated that the power level was still climbing, on its way to disaster. The assistant at the controls screamed over the phone for the supervisor to do something about the air pressure, so that the control rods would drop and stop the chain reaction. There was no way to do this. The combination of errors had snowballed into an uncorrectable situation.

Exactly forty-four seconds after the accidental pushing of button number 1, a plant physicist realized that the only thing left to do was to dump the heavy water from the reactor, and thus cut off the fission process. There were thousands of barrels of heavy water in the reactor—each barrel worth more than a

Cadillac. But it was the only option. The physicist reached over and slammed the dump switch.

It took several seconds to see what would happen. The power seemed to drop, but almost immediately another hazard loomed up: The whole sealed reactor vessel might collapse from the vacuum formed by the dumping. The operation was halted, then cautiously resumed. A sigh of relief went up when the instruments went back to normal, about thirty seconds after the dumping had begun.

But the disaster was far from over. Someone looked through an open basement door, and saw tons of water rushing out of the reactor, flooding the basement area. The supervisor and his assistant rushed with a bucket, carefully handled at a distance, to take a sample. A quick test showed it to be ordinary light water, but highly radioactive.

Then, four minutes after button number 1 had been accidentally pushed, a dull rumble was heard. The huge, four-ton lid on the reactor vessel, called a gasholder, rose in the air. A spurt of water gushed out through the top of the reactor, spilling over the building floor. Radiation alarms went off, and the sensor near the steam fan showed lethal doses of radiation escaping.

From a nearby building, a frantic phone call came in to report that the readings in the atmosphere in that vicinity were far above scale. The caller requested what was called an emergency stay-in procedure. A siren sounded, alterting all personnel to proceed to the nearest building, close all windows and doors, and to use the telephone only in an emergency.

With radioactive contamination rising rapidly by the control room door—not yet deadly, but a warning—gas masks were issued to the crew at the control panel. But critical and urgent discussions were impossible through the masks, and the crew was forced to retreat to another, less contaminated building.

At 3:45 P.M. the project director and the radiation hazards

control director gave the order to evacuate the entire installation, including buildings and grounds. All but the essential crew followed the prepared procedure: hurrying in an orderly manner to the gates, holding a handkerchief over the nose and mouth.

Meanwhile, the reactor crew, removing gas masks only for the purpose of discussion, stood by helplessly as more than a million gallons of highly radioactive water flooded the basement of the reactor building. If they tried to stop the water, the deadly melted fuel in the reactor might catch fire and make the situation even more disastrous. The flooded water contained ten times the amount of long-life radioactivity that there was in the entire world in 1940. But gradually, in several hours, the reactor tamed down.

Because of the small size of the reactor and its remote location away from cities, the damage was minimal. The painful process of decontamination was begun cautiously the next day: scrubbing every square inch of surface in the eight-story-high NRX building by mop and sponge soaked in detergent; burial of all the soiled cleaning materials; the donning of hot, sweaty plastic suits and Scott-Pak respirator masks; the hurried construction of a pipeline to a sandy valley over a mile away to dump the million gallons of radioactive water so that it wouldn't contaminate the nearby Ottawa River. The swollen and melted fuel—a lethal source of radiation from the melted uranium—had to be kept cool by connecting water hoses to each rod.

The amount of radiation that each man would receive during the decontamination job had to be rationed carefully. Radiation doses are cumulative; the time of exposure must be limited. Skilled operators were spared in the early clean-up stages to prevent them from absorbing so much radiation they would be unavailable later. Personnel from other departments were solicited for the risky job and trained on a special mock-up of the reactor. The simple job of removing one flange and inserting a diaphragm in the coolant pipes took twenty men, clad in full protective suits and gas masks, and working in carefully planned relays. The radiation at this location was so hot that only moments could be spent there.

There was evidence of a hydrogen-oxygen explosion inside the reactor, along with the melted uranium that had riddled and scarred the guts of the core. But the general consensus was that Chalk River and the men working on the NRX reactor were lucky. There was no explosion outside the reactor. The uranium melting was contained.

As the melted fuel and broken steel were finally bagged, and dragged away for burial, there was measurable relief in the community of Chalk River. Especially when a post-mortem review revealed that if one more control rod had jammed, the increase in the fission products released into the air could have wiped out the bedroom village of Deep River and beyond.

From the accident came lessons for the entire international nuclear community. One defect leads to another—and another. One human error does the same. Most are unpredictable. This snowballing throws off all the carefully calculated engineering probability studies as to the chances of a major accident. Most single errors might be controlled. It is the errors compounding the errors that no computer can foresee. Murphy's Law: "If anything can go wrong, it will."—the bugaboo of all engineers—would be stalking the nuclear energy field, as everywhere else.

The awesome consequences of a full-scale nuclear power plant accident automatically demand that the design, construction, equipment, and the men behind it all must be infallible. The Chalk River accident produced a major question: Was infallibility possible?

With an untested breeder reactor, twenty times more powerful than any that had been built before, situated in the middle of a huge population center, this was the question that Walker Cisler and his skilled team of engineers at Lagoona Beach would have to answer.

TWO

Like everybody else in the business, Walker Cisler was painfully aware of the lessons to be learned from the Chalk River accident. He would have to keep in mind that the breeder reactor his group was planning was not only more dangerous and sensitive than the Canadian heavy-water reactor, but so many times more powerful that the consequences of an accident could be many orders of magnitude more disastrous.

But problem-solving was a built-in characteristic of the dynamic Cisler, and he began gathering men around him who were confident that they could conquer any sort of safety problem. To accomplish this they would have to examine every possibility of an accident in infinite detail. They would have to measure any conceivable combination where things might go wrong, and make sure the design of the reactor would provide for it. Even the most farfetched possibilities would be considered—and the means of protecting against them flawlessly worked out.

All through 1953 and into 1954, the plans moved slowly. It was no easy job to gather a staff that would have to meet the criterion of infallibility, but the group that joined Cisler in his project shared common motivations. They knew well the growing need for new sources of power and energy—long before the public ever gave any thought to the situation. And as men of enterprise, they were confident and determined that they could best meet this need through the concept of the nuclear breeder, the unique machine that could make its own fuel.

They had pride in both their vision and craftsmanship, and they were not without vanity. They also had the necessary incentive. Most important, there was a challenge here, and if the question of safety could be solved, meeting the challenge would not only eventually bring profit, but immense benefits for society as well. Although there was still no official government sanction for private industry to go ahead with a nuclear power program, Cisler and his team were making sure they would be ready for it when it came.

When Eisenhower signed the amended Atomic Energy Act into law at the end of August, 1954, it was the signal to move full speed ahead. For the first time, atomic facilities could be owned by private companies. Cisler had already gathered $2.5 million for the basic research on the reactor, much of the research already having received tacit approval from the AEC through an industrial committee set up by the commission some years before.

Further impetus for Cisler's project was President Eisenhower's appointment of Lewis Strauss, an investment banker who had long been active in government affairs, as the new chairman of the Atomic Energy Commission. Strauss was a strong supporter of getting nuclear power out of the AEC and into the private sector. He was a brittle and controversial figure, but like Cisler he had the capacity for bulldozing a project through to the finish.

One of the first problems for Strauss to determine was which type of reactor was practical for private industry. In addition to the breeder, the type of reactor that Admiral Rickover was successfully developing for submarines also seemed logical. It was called the light-water reactor. The basic principle was the same for both: the splitting of atoms in the uranium core to produce heat, which in turn would produce steam to turn the generators. But only the breeder could capture the excess neutrons to create new fuel, and this is what made it so appealing to the imagination. Under Strauss's plan, public utilities would be encouraged to go ahead with both. The light-water reactor, being more predictable and less risky, would be taking the lead in popularity among those utilities who feared the uncertain, untested, unexplored quirks of the breeder.

Both types of reactors had safety problems. The fuel in a light-water reactor cannot form a critical mass. But the high pressures necessary for the light-water reactor made it subject to critical failures. For example, the effect of long-term radiation on the metals of the fuel rods was unpredictable. Or under accident conditions, the molten core could fall into the water. The 5,000°F core could then cause a violent steam explosion and breach the containment, spewing radioactivity. There was also the possibility of earthquake, sabotage, or human error.

The breeder reactor had special problems. Besides producing plutonium, its core could turn into a critical mass. It also was cooled by liquid sodium, a thick viscous fluid that is subject to special risks. In contact with water or air, the sodium could explode and flash into fire. In case of disaster, events would occur one thousand times faster than with the light-water models.

Cisler was confident all these problems could be solved. He decided to forge ahead with his plans for the breeder reactor, while other utilities in other parts of the country generally favored the idea of the light-water reactor. Of major importance was the offer of governmental aid. All the utilities involved in developing either type of reactor would be receiving considerable help. Design research would be made freely available, and uranium fuel would be furnished at a fraction of the cost that the

AEC spent to produce it. And since the plutonium produced by the breeder could be sold back to the government, as the reactor continued to produce more fuel than it consumed, the breeder reactor presented a more attractive long-range picture.

It was so appealing that more blue-chip utilities joined in with Cisler's breeder project. Dow Chemical, however, withdrew to concentrate on the chemical aspects of the nuclear field. At a later date, Ford and General Motors joined in, too. The new consortium was named the Atomic Power Development Associates, with Walker Cisler as president. They would be facing endless meetings with the AEC as they began to explore the potential of taming the mercurial atom.

The project staff was as aware of safety as anyone else, and they would have to be considering both major and minor dangers of a breeder reactor. The possibility of a runaway meltdown is the most critical problem that any reactor, breeder or otherwise, must face. When nuclear fuel, usually uranium, melts like a candle into a waxy, drippy mass, it can become unpredictable. It might melt down through the bottom of the reactor vessel in what engineers call the "China Syndrome"—the molten mass of uranium heading down through the earth toward China. If it forms into a thick mass, it is possible for it to cause either a chemical or a small nuclear explosion that might breach the containment building. In turn, this could release a silent, odorless, tasteless, colorless cloud of radioactive gases and particles into the atmosphere. The resulting plume could contaminate, kill, and injure as much or more than dozens, scores, or hundreds of Hiroshima-type bombs—depending upon the size of the reactor and the amount of fission products built up in the fuel as it becomes depleted.

No scientist, even the strongest supporter of nuclear power, disagreed that, if the unthinkable happened, there would be a massive and unprecedented catastrophe.

None of the men in the AEC or private industry took this

responsibility lightly. They were as concerned as anyone about their homes, their families, and their duty to the public.

When Cisler's staff met with AEC officials at the Detroit Edison offices on November 10, 1954, in a guarded, classified meeting, the safety issue was on everyone's mind. Present was Walter Zinn, father of the EBR-I which had produced the first token atomic electrical power, and Hans Bethe, the Nobel laureate and physicist from Cornell, who was acting as a consultant for Cisler's group.

The EBR-I was of critical importance to Cisler's project, because it was a Tinker-Toy model of what would later be built at Lagoona Beach. The enriched Uranium-235 fuel would be almost the same for both reactors. If a runaway meltdown should develop, and the fuel should reassemble itself at the bottom of the reactor, it was entirely possible for a breeder reactor like EBR-1, or the breeder planned for Michigan, to turn itself into a mass similar to that of a nuclear bomb, not with the same explosive power, but still a significant contamination potential if the containment were breached.

There were various names for this condition. Some called it a "superprompt critical power excursion." Others merely referred to it as "prompt critical." Whatever it was called, it could create a chain reaction that might spin out of control, and then nothing could be done to stop it. The major consequence would very likely be a breach of containment, and the spreading of radioactive fallout.

Zinn and Bethe agreed about what would happen if the central section of the reactor core were to melt and run down the tubes, and both made it clear that this could be a disastrous event. If the fuel rods melted or warped, no one could predict what kind of configuration might result. If it were compact, Zinn felt, it could "disassemble the machine." In plain language, this meant a nuclear explosion.

Although these possibilities were frankly admitted, the men

at the meeting were confident that engineering know-how could make sure that this unthinkable possibility would never happen. But the breeder, they felt, was the ultimate answer to saving the world from a pending energy crisis. They were convinced that their skills could conquer the dangers and bring the benefits. Cisler was passionate in his beliefs. "The breeder reactor," he said to his colleagues, "would continuously produce amounts of fissionable material in excess of that consumed. Thus breeder reactors would augment rather than consume the world's supply of fissionable materials."

The major question before these leading physicists and engineers was never quite fully answered at the meeting: Was it safe to place a developmental breeder reactor halfway between Detroit and Toledo, within thirty miles of each city, in the heart of an area embracing a population of over four million?

In spite of the controversy that was to grow over the years, the scientists who met at Detroit Edison on that November day in 1954 were both inspired and determined to put the fast-flying neutrons of Uranium-235 to work for the benefit of society. Although their critics were saying that they were making an unacceptable value judgment for the public, they were dedicated to the task of building a reactor that would not threaten their families, themselves, or the population. Their foreknowledge of the potential dangers and their skill in dealing with them would be their protective armor.

Shortly after the meeting, Walker Cisler announced that ground breaking for the new fast breeder would definitely take place within the next five years. But Cisler's optimism was not shared by another group absolutely necessary for the building of the reactor: the insurance industry. With the potential for a catastrophe created by any sort of reactor, whether it be fast breeder or light-water, it would be suicide to begin operations without insurance.

The insurance men were chary and timid. They looked at the nuclear power plants on the drawing boards, Cisler's among them, with a fishy eye. The potential danger to the public appeared so incredibly great that not a single insurance company was interested in taking the gamble, not even Lloyd's of London, the greatest of risk takers.

Chalk River had proven that accidents could happen. There was still no realistic estimate as to exactly how many people would be killed, maimed, or come down with leukemia if an "energy release" hit a populated area. To insure enriched uranium or plutonium could hardly be defined as a conventional risk. As the most toxic substance known to man, it has been estimated that even 1/30,000,000th of an ounce of plutonium could bring on cancer if inhaled. And what worried the insurers most were the plans being formed by the AEC that called for the creation of several hundred thousand *pounds* of plutonium by the end of the century, all of it possessing a hazardous life span of a minimum of 480,000 years.

The insurance men were realists. And they were in somewhat of a squeeze. They did not want the government to take over the insurance function in this new, fast-moving industry. Yet they had no firm data on which to estimate the risk they would have to assume.

It was becoming a serious roadblock. Westinghouse Electric, which became part of the elaborate conglomerate, flatly stated that it would not be able to go ahead with the construction of any atomic power plants unless it could get protection against the enormous losses that would result from a major nuclear accident. Westinghouse was joined by Con Edison, the New York utility, which stated that it would not think of operating its planned new light-water reactor if there was no insurance available.

Charged with the strange and conflicting responsibility of both regulating and promoting the fledgling nuclear industry, the Joint Committee on Atomic Energy of Congress and the AEC were aware that, unless action was taken to provide insurance protection, no atomic plants whatever would be built or operated. Some proposals were being made whereby the government might supplement whatever token amount the insurance industry was willing to risk. One idea in the wind was a proposal by Congressman Melvin Price and Senator Clinton Anderson for a plan whereby the maximum coverage would total only $560 million. Of this total the industry was required to obtain as much insurance as the private insurance pool would provide and the federal government would provide the rest of the insurance up to a maximum amount of $500 million. Since the private insurance companies were willing to put up only $65 million—a drop in the bucket compared to the damages that might result from a meltdown—the federal government's share was $495 million. Critics of the proposal pointed out that, not only would the public taxpayer be paying for private industry's insurance, but that the ceiling limit might leave thousands of victims unindemnified.

Cisler realized that all the engineering know-how he could muster could not get around this obstacle. He began pushing "war-risk insurance" from the government—despite his stand against government encroachment. He had not yet made a formal proposal to the AEC to build the reactor. However, the research and development work had moved ahead slowly but persistently, and the formal application would not be far off.

A good many scientists did not share Cisler's confidence in the feasibility of building a commercial breeder reactor in a heavily populated area. Among them was Dr. George L. Weil, a former research associate of the late Dr. Enrico Fermi at Columbia—the man for whom Cisler's reactor would be named. George Weil had been a member of the team of scientists that had brought about the first sustained nuclear chain reaction underneath Stagg Stadium in Chicago for the wartime Manhattan Project. He had recently resigned as the chief of the reactor branch of the division of research for the AEC, and become a private consultant to companies interested in going into atomic energy under the new government incentives for so doing.

In April, 1955, several months after the secret meeting at Detroit Edison, Weil was asked confidentially by one of his clients, a large utility company, whether it should join in with Walker Cisler's pioneering effort. Weil gave the question serious thought. His long experience, from the first splitting of the atom at Stagg Stadium to his high responsibilities with the AEC, had carried him through a painful process of reexamination, and he had become specifically concerned with what was happening in the growth of radioactive poisons. He was not at all happy about it.

When he had started working with the Manhattan Project during World War II, radium had been the only source of the poisons. The entire amount in the world totaled a mere 1,000 curies of radium. The thinnest chip of radium is deadly. It must be kept in a thick lead container and handled by remote control.

But with the nuclear power plants being planned for the future, it would be possible for a *single* nuclear power plant to contain radioactive materials equal to *20 billion curies* of radium. With future projections showing hundreds of large nuclear plants across the country over the next fifty years, Weil began asking himself: "What kind of sword of Damocles will be hanging over our heads now and in the future?"

Because of this, Weil could not go along with the idea of his client joining the Cisler power plant consortium. On April 26, 1955, he wrote a letter to them that said:

". . . it is my opinion that the time is not now ripe for the construction of large scale developmental fast breeder reactors [the word *fast* referred to the high-speed neutrons]. The opinion is based on the following considerations:

"1. The technology of the fast breeder reactors is in the early stages of development.

"2. There are many difficult technical and engineering problems which must be solved before commercially feasible fast breeder reactors can be constructed.

"3. The solution of these problems will involve a long and costly program."

But there was also another very practical and subtle thought underlying Weil's thinking: If there were to be only one major nuclear plant accident (and even the AEC would agree that this was entirely possible), the public outrage would be so great because of the catastrophic loss of life and injuries, that it would immediately spell the end of nuclear power. Thus, the billions of dollars invested would be wiped out—to say nothing of the estimated billions of dollars damage caused by the accident. Was this precarious scaffold a financially sound platform to build on?

Weil's letter went on to say that only small, pilot-scale reactors should be attempted; the chance of success of the Fermi reactor was small indeed. Because he was thinking in terms of his client's benefit-risk outlook, his recommendation was based purely on cold, profit-and-loss practicality. His client agreed, and declined to join in the project.

In the months that followed the first classified meeting about the Cisler plans, Alfred Amorosi was named technical director of the project. His scientific design and research team began examining every possible doubt about the safety of the fast breeder. They were fortunate to have the experimental reactor EBR-I out in Idaho Falls as a pilot plant. They would be depending on it to provide test information about the safety of their project.

Although the EBR-I was of such low power that it could in no way act as a realistic blueprint for the full power commercial reactor contemplated by the Detroit Edison group, it could furnish a ground for experiments to forecast stability and check weaknesses. Along with other experimental reactors at the AEC test station in Idaho Falls, the EBR-I also could supply information on what might happen if there were a meltdown of the commercial reactor's core.

Amorosi and his staff directed their attention to two special danger areas in the design of the breeder reactor. One concerned the ability of the reactor to decrease in reactivity as its temperature rose—an important safety concern when dealing with split-second timing. For maximum safety, the reactor should have what is called a "negative temperature coefficient" to help control any sudden power surge by keeping the power and temperature down.

Another subtle characteristic that could spell the difference between safety and disaster was what was termed the "Doppler effect." If this was negative, it was desirable, because it slowed down the fissioning. If it was positive, it was dangerous, because it could increase an already risky power surge. The control of a reactor was often as delicate as trying to adjust the hot water in a hotel shower at breakfast time.

Both of these problems were carefully examined in Washington on June 30, 1955, at another closed session of the Advisory Committee on Reactor Safeguards. This was a highly prestigious panel of technical and scientific specialists formed by the AEC to screen all plans for nuclear power plants. The panel was intensely interested in what Amorosi's computations showed as far as these two built-in safety devices were concerned.

Amorosi's figures showed that he was confident that the "negative temperature coefficient" would serve as an adequate brake in case of an unexpected temperature rise. His computations on the Doppler effect, however, showed that it would be positive—meaning that it would add to the problems if the reactor started to go out of control.

The specialists on the Advisory Committee took a long hard look at these points at the Washington meeting. The EBR-I, similar in concept to the Detroit breeder reactor, had not been a perfect machine, by a long shot. It had been showing its own problems with automatic built-in safety checks. It was also having an oscillation problem, observers having noted wide swings of unpredictable splitting of the atoms.

The panel going over Amorosi's planning and progress report in the closed session was somewhat ambivalent about the entire picture. For example, shouldn't all the problems with the EBR-I be straightened out before basing new plans on its design? Or, if the EBR-I was showing the dangerous "positive temperature coefficient," why should they expect the larger commercial reactor not to have it? What would happen to the stability of the reactor after plutonium was formed? What if excessive fuel was loaded into the reactor? What effect would the hot liquid sodium rushing through the reactor have on these safety checks? None of the Detroit Edison theories had been checked experimentally. How could this be done safely?

Amorosi and the Detroit group contended at the meeting that all this could be done after the reactor had been built at Lagoona Beach. But several members of the Advisory Committee did not think that was proper or safe.

The results of the June, 1955, meeting were ambiguous. Later, in executive session, the Advisory Committee went along with the calculations Amorosi had presented, but with very clear reservations: "It must be recognized that the assumptions on which these calculations are based have *not* been established experimentally," their closed session report read, "*and must be so before the operation of such a reactor could possibly be recommended for a site so close to* a populated area." [Italics added.] Then the report continued: "If, as seems likely, this program includes start-up and low-power transient experiments with the reactor itself at the Michigan site before start-up, it must be established that such experiments in themselves cannot lead to a containment-breaching incident."

Among those who were determined that no such thing would ever happen was young Walter McCarthy, head of nuclear engineering for the Michigan project. Like Cisler, McCarthy was a Cornell engineer, and he was a doer. Poised, confident, and alert, he was a lanky Irishman who had tackled the intricacies of nuclear reactors at Oak Ridge. He was also a thinker, and he spent long weeks analyzing just what hazards the fast breeder might develop.

McCarthy's safety studies were infinitely complex and exacting. Among other things, he was concerned about the possibility of "an explosive energy release" occurring in the core, which of course would be a bloodcurdling event of unknown proportions. "Since the concentration and amount of fissionable material in a fast power reactor is sufficient to produce a number of critical masses if arranged compactly," he wrote in a cooperative study with an AEC physicist, "a number of investigations have been made to consider the redistribution and reassembly problem."

McCarthy was talking about the enormous danger of the fuel in the reactor melting in microseconds, and rearranging itself into an unpredictable shape that could explode, breach the containment, and spew the massive amounts of radioactive poisons into the air. Calculations up to this time had shown that a severe meltdown accident in a commercial-sized nuclear breeder plant could create an explosion equal to a thousand or more pounds of TNT. McCarthy's analysis indicated that there were many unknowns, few answers to critical questions, many problems unsolved, and very little experimental information in other important areas. Speaking of the unknown area of the explosion possibility in a large nuclear plant, McCarthy's paper concluded: "The possibility that only a portion of such a reactor melts, undergoes a relatively mild explosion which acts to compress other parts of the core extremely rapidly, thus instigating a much larger energy release [explosion], needs further investigation."

McCarthy had noted that the area of meltdown investigation remained one of the central problems in fast breeder safety studies. So much could occur so quickly that a reactor could be destroyed before the control system would have time to react. He also recognized that many of the calculations regarding the new liquid sodium coolant, so necessary to prevent an accident, were imperfect.

These were the sorts of problems that the engineers and scientists working with Cisler were facing. But McCarthy had the confidence that he and his fellow crew members could overcome them. It was a challenge, and McCarthy reveled in it. He and other staff members made frequent trips to the sprawling, lonely AEC station at Idaho Falls to check out their calculations in a series of mock-ups on the EBR-I breeder and other small-scale reactors in the area.

Meanwhile, the insurance companies remained unconvinced that a commercial nuclear reactor near a large population center was an insurable risk. The entire pooled insurance industry was not willing to offer any measurable kind of protection beyond the $65 million token coverage. And what was to happen with the EBR-I reactor in Idaho Falls in November of 1955 did little to give the insurance companies confidence.

One of the problems the AEC scientists had worried about earlier was the sudden changes in fuel temperature at EBR-I. A decision was made to run some tests and find out what was causing the situation.

The central core of the EBR-I was shaped like a hexagon.

Packed tightly into it were some two hundred stainless steel tubes called cladding. They looked very much like curtain rods, less than half an inch in diameter. Inside each were two pellets that looked like miniature Tootsie Rolls which were made of highly enriched Uranium-235. Around the central hexagon core were stacked the thicker rods of what was called the inner blanket. These were made of raw Uranium-238. They were bombarded by the leaking neutrons from the core and converted into plutonium.

In an outer circle around both the core and the inner blanket was an air-cooled outer blanket of Uranium-238 bricks. Down through this wall of bricks ran the control rods, twelve of them. When they were withdrawn out of the pile, the splitting of the atoms would begin; if inserted all the way down, the splitting stopped. As a further safety measure, the entire outer blanket could be dropped down out of the fission action, providing for an emergency shutdown of the reactor.

The EBR-I reactor would be started at very low power—just a few watts. Then the power would be increased. Special instruments were attached to key parts of the reactor mechanism to keep track of the temperatures. They were going to be allowed to rise to over 900°F., not too many degrees under the melting point of the fuel. It would be a tricky operation. The control room staff would have to be alert. And it was most essential that the operator at the control be ready for an extremely rapid shutdown at exactly the right moment. However, there was an experienced, knowledgeable staff on hand, and there was little concern about their capacity to handle the situation.

The experiment began normally at the very low-power level of 11 watts—barely equal to the smallest of light bulbs. The motor-driven control rods were slowly withdrawn out of the core, and the fission reactivity increased up to 50 watts.

As the power level moved up to 500 watts, there were confusing temperature readings among the special and standard instruments. It became apparent that the reactor might be on its way to an excursion—a sudden rapid rise in the power level.

The plans, carefully laid out in advance, had prepared for this sort of emergency. The scientist in charge was to give a verbal command to the operator at the control panel. He, in turn, was to immediately trip the fast-action control rods. The slower, motor-driven rods could not handle the shutdown in time.

With the temperature and power now definitely rising out of control, the head scientist gave the command for the operator to trip the fast control rods. They were dealing now with a one-second leeway. By mistake, the operator hit the button for the slower, motor-driven control rods.

The power continued to rise, doubling every two-tenths of a second. The chief scientist then realized what had happened. He reached over and hit the rapid shutoff button. The safety rods responded—but the power began increasing again. The instruments went off scale.

Quickly, the scram button was pushed, which dropped the Uranium-238 outer blanket to reduce the fission reactivity. But within fifteen minutes, radioactivity was registered in a cooling system some distance from the reactor, and in the ventilation exhaust ducts.

The building was immediately evacuated. A health-physics team moved in. They found low-level contamination by radioactive fission gases, but fortunately no one was hurt.

Nearly half of the core of the small reactor had melted, foaming and frothing as it did so. The temperatures had reached over 2000°F.—much more than the melting point of the fuel and stainless steel cladding. The liquid sodium coolant had boiled over, pushing the uranium outward from the center of the core and blocking coolant channels. Partly melted rods dropped into a molten mass below the core, forming what is known as a eutectic mixture. The mass finally chilled and froze. Fortunately, it did not turn itself into a critical mass—an event that would have made an explosion a strong possibility.

When the EBR-I crew began licking its wounds after the excursion, they made several discoveries. One was that a great many questions about the unsafe characteristics of the breeder reactor remained unanswered. Another was that, with the power of the reactor doubling every tenth of a second, there was no margin for error. Still another was that the fuel rods had probably bowed out of shape before the accident, and had been the major cause of it.

But it was also obvious that, regardless of skill and know-how, there had been both human and material deficiencies. And if it had not been for the emergency dropping of the blanket (the blanket "cup," it was called), the reactor would have been one-half second away from either a nuclear runaway or a "meltdown crash-down." In this condition, the molten center of the core would freeze in a tightly compact mass, followed by the upper part of the core crashing down on it. The possibility of a low-order nuclear explosion in such a case was real.

The EBR-I accident had serious implications for Cisler's team. If the core of the EBR-I had been plutonium, there was little question in the minds of experts that a disaster could have occurred, even at the remote Idaho location. If the EBR-I had been of the larger commercial size, the results of the accident would certainly have been much more severe.

Because this reactor was critical to the design of the Detroit reactor, there were ominous implications for the future. It was obvious from the accident that human error could not be ruled out. Yet with a commercial reactor in a populated area, it somehow had to be. The words from the carefully prepared scientific study of the accident by MIT scientist T. J. Thompson would remain in the minds of the Cisler group for a long time to come: "It is still a fact that people and equipment are subject to failure. . . ."

The study of the accident left much conjecture, many questions unanswered, and considerable doubt about what actually happened. And the burning question also remained: Would something like this happen in other reactors?

Several members of the Joint Committee on Atomic Energy were most disturbed, because, like the Fermi project at Lagoona Beach, the EBR-I was a breeder reactor. It could well reflect the type of problem that might have to be faced with commercial reactors in the near future.

The accident was also sharply disturbing to the insurance companies. They would ultimately be the ones to determine whether any reactors would be built or not—unless the government came through with the taxpayer-financed insurance. The idea of "war-risk" insurance seemed totally out of character with a peacetime venture. Yet there seemed to be no other way out. The accident also punctuated the idea that if a commercial reactor was so devastatingly threatening as to require war-risk insurance, with damages equal to or greater than the ravages of war, the risks would seem to be too enormous to contemplate.

Henry Young, a vice-president of Liberty Mutual Insurance Company, was particularly vexed by the situation. No one as yet had come up with any realistic figures about the highest amount of damages that might be leveled on the public through a major nuclear accident. To Young, the catastrophe hazard appeared to be many times as great as anything previously known in industry. And he voiced his fears to the Joint Committee on Atomic Energy: "We have heard estimates of catastrophe running not merely into millions or tens of millions but into hundreds of millions and billions of dollars. It is a reasonable question as to whether a hazard of this magnitude should be permitted, if it actually exists. Obviously there is no principle of insurance that can be applied to a single location where the potential loss approaches such astronomical proportions." Then he added: "Even if insurance could be found, *there is a serious question whether the amount of damage to persons and property would be worth the possible benefit accruing from atomic development.*" [Italics added.]

So the nagging question was: If nuclear power plants were safe, why wouldn't the insurance companies back them up? Or as another critic put it: "If the insurance companies don't believe the AEC's promise that there is little chance of an accident, why should the people living in the same region believe it?"

If this attitude were to prevail, Cisler's dream would never come true. But neither he nor the AEC nor the congressional Joint Committee were about to back off in spite of the Chalk River and EBR-I accidents—or the seemingly impassable roadblock as far as insurance was concerned.

THREE

By the first week in January, 1956, Cisler had officially applied for a construction permit at the Lagoona Beach site and was waiting for permission from the AEC to proceed. He formed another non-profit combine called the Power Reactor Development Company to carry out the actual construction and operation of the reactor. The new combine consisted mainly of companies that had joined him purely for the design work of the reactor. The earlier combine had been called the Atomic Power Development Associates, and the two units became known as the APDA and the PRDC, joining the proliferation of acronyms in the atomic energy field that were growing like a spilled box of Scrabble letters. What counted was that the two organizations were really Walter Cisler, and the new reactor was to be named the Enrico Fermi Atomic Power Plant. The first cost estimate for the project came to $40 million.

The preparation of the license application for the new Fermi plant was a massive job. It amounted to many thick, bound volumes of hundreds of pages covering every aspect of the operation from design and fabrication to an analysis of the expected hazards. The philosophy laid down by Cisler was to design the plant so that no credible malfunction or accident could release any of the deadly radioactive fission products from the reactor. Further, if the incredible should happen, there must be no way that the tightly sealed containment building could be breached by an explosion or a dreaded sodium-air reaction that could eat up all the oxygen and collapse the building.

The safety hazards section of the license application was prepared with meticulous care by Cisler's staff. Every conceivable type of accident was spelled out in detail, and the ways of controlling it assessed. One possibility concerned a surge of reactivity of the chain reaction during the operation or during the loading of the fuel. Another was the fast reassembly of the material in the core during a meltdown when the uncontrolled fuel would pile up dangerously. Either could lead to a nuclear runaway or explosion.

These possible accidents were divided between those which were considered "credible," and those which were thought to be "hypothetical." The "hypothetical" group was defined as being so improbable that they were incredible. In either classification, the license application examined such possibilities as primary sodium system leaks (bringing fire and explosion on contact with air or water), loss of plant electric supply (with the loss of coolant and a runaway meltdown), the dropping of a core subassembly during refueling (another uncontrolled meltdown possibility), the failure of the safety rods to fall (another runaway situation), and the ultimate meltdown accident—one that the containment shell failed to hold.

Because there was no experience to work on regarding many of the possible ways a runaway meltdown might go, much of the theory had to depend on guesswork. It was like trying to predict how the logs in a fireplace might fall as they burned. One of the most feared conditions would be that of a secondary accident after the first part of the meltdown had chilled and frozen. The slightest disturbance of a pile of melted fuel could cause unpredictable havoc. Yet the written application was forced to note that "no experimental data" were available on this.

Because of the importance of the containment shell—the last barrier between a meltdown and the public—the plans for its construction were elaborate. Hans Bethe had calculated that the containment would have to stand the force of a nuclear explosion equal to that of five hundred pounds of TNT, although there were other nuclear scientists who claimed the explosion could be much greater. Some independent studies indicated it could be up to twenty times that force.

Two Fermi consultants studied what might theoretically happen to a containment building that would have to hold a blast equal to five hundred pounds of TNT. They calculated that the blast wave would deform the wall permanently, but that the steel and concrete would keep the deformations within tolerable limits. They estimated that the operating floor would crack, but not endanger the containment vessel. They were uncertain whether the materials inside the reactor itself, including the sodium piping and the reactor vessel, would fracture and form missiles. However, if such missiles were formed, the scientists did not think they would break through the containment walls.

The missile problem had to be considered from both inside and out of the big reactor dome. The velocity of a tornado had never been measured, but it was known that it could hurl huge objects with deadly power. It was thought that a 35-foot telephone pole weighing 1,600 pounds, going 150 miles an hour, could be slammed against a nuclear power plant building by a tornado, and that the containment shell should be designed to withstand this. When it was discovered that it was practically impossible to design for such a contingency, the criteria were relaxed so that the shell would only have to protect against a four-inch by twelve-inch wooden plank.

One exterior hazard that still hangs over every nuclear

plant is the possibility of heavy modern aircraft falling into it. Because the probabilities would be so small, this factor was generally dismissed. Other considerations were earthquakes and floods which would be equally dangerous.

Armed with the written application, estimates, and blueprints, the men of the Fermi project met to present their case to the Advisory Committee on Reactor Safeguards in early June of 1956. The decision to permit Cisler to go ahead with the actual construction of the reactor rested heavily on this meeting. He gathered his best men around him, among them Amorosi, McCarthy, and Hans Bethe, and the group met at the AEC's Argonne National Laboratories, outside of Chicago. The chairman of the Advisory Committee was the highly regarded atomic scientist C. Rogers McCullough of Monsanto Chemical. He headed the group of a dozen nuclear experts, including Dr. Edward Teller.

The meeting—and the eventual approval of the construction permit for the Fermi reactor—centered around one very important question brought up by the AEC's division of civilian application. "Is there sufficient information to allow the AEC to state with reasonable assurance that a reactor of the type under consideration can be constructed and operated at the site selected without undue risk to the health and safety of the public?"

There were also a lot of ancillary questions: Would the reactor be stable enough so that there would be no risk of a meltdown? Would a disastrous nuclear explosion result from the manner in which the fuel arranged itself in a meltdown? Would the design of the containment building really hold the radioactive debris safely? Could experiments be carried out at Lagoona Beach before a full-power operation, without the experiments themselves being too much of a safety risk? Could even the start-up tests at low power be conducted safely? How safe would the Fermi reactor be after substantial amounts of plutonium had built up in the core blanket, where it would be bred?

These were not easy questions. But they were vital as far as the public was concerned. On everybody's mind was the EBR-I accident. Walter Zinn talked about it at the meeting, and showed pictures of the melted, twisted core. There was still a big question mark as to why the accident had happened. The destruction of this reactor core had made many tests for the Fermi plant impossible to carry out. A new EBR-II was being constructed, but this would be completed too late to serve as a model if the Fermi plant was already under construction, as Cisler hoped it would be. But also, neither the EBR-I nor the EBR-II could really test what would happen to the bigger commercial reactor once it went into operation. Other plans for adequate checking out of the Fermi design were either too premature to be of use, or simply inadequate.

One suggestion was that a few fuel rods of enriched Uranium-235 could be assembled with a Uranium-238 blanket around them, then placed in a pot of liquid sodium and brought to a meltdown condition for testing. But even a simple assembly like this would take well over a year to create. Elaborate safety precautions would be needed, even with the smallest collection of enriched uranium.

Cisler's group, pushing for approval, pointed out that the Naval Ordnance Laboratory had estimated that the design would hold an explosion of five hundred pounds of TNT. Some of the arguments for approval, however, seemed weak in the light of public safety in the Detroit area. The start-up test program at the Lagoona Beach site was going to be relied on to check out whether or not an unpredictable surge of power or temperature would be a characteristic of the new Fermi reactor. The Cisler team argued that, since the tests would be at a low-power level, there would be less chance of runaway meltdown. The general attitude of the Advisory Committee had been that there should be no chance whatsoever of a runaway meltdown.

Another point that was discussed had ominous implications.

The critical testing of the reactor would be carried out only when the wind direction was favorable for "minimum population exposure" in the event of a radiation accident. This implied a suggestion of insecurity in the safeguards that were supposed to be so airtight. But beyond that, the wind direction on the Ohio shore of Lake Erie was notoriously fickle. And regardless of the wind direction, there would be a considerable number of people exposed, especially the several hundred homes in the Stony Point area.

Left unanswered was what would happen to the stability of the Fermi reactor after the plutonium had built up, and after the thin rods of fuel had been battered by constant radiation. Also left up in the air was what would happen if the fuel melted, and the reactor was still unable to be kept under control. The secondary aspects of an accident were often more terrifying than the accident itself.

Cisler and the Fermi group left the meeting in a mist of uncertainty, as the Advisory Committee continued in executive session. The latter group was facing a rather awesome responsibility to the public. But as an independent arm of the AEC it was also faced with the policy directive from Chairman Lewis Strauss to push hard for the development of commercial atomic power. This conflict was mounting constantly, as other utilities began cautiously to get in on the tempting new source of energy. The anomaly lay in the continuing position of the AEC and the Joint Committee on Atomic Energy to regulate on one hand and promote on the other. The Advisory Committee on Reactor Safeguards was the only real regulation force. But, inasmuch as it actually reported to the AEC, its independence was in question.

After Cisler's group had gone, the Advisory Committee argued long and ponderously about the wisdom of giving a go-ahead to the Fermi project. The twelve experts spent two more days trying to come to a yes-or-no answer to the AEC's civilian application division's question: Was there enough information to say that the Fermi plant could be built without undue risk to the health and safety of the public?

Special studies on the outside limits of danger from a nuclear power plant were already underway. Congress had asked the AEC to estimate in exact terms how many people would be killed and maimed, and what sort of property damage would result from a hypothetical major reactor accident near a large city somewhere in the U.S. To that end, the AEC had commissioned a study by a group of scientists and engineers at the Brookhaven National Laboratory on Long Island. This study would come to be known as the WASH-740 report. The results of the study would have an important impact not only on the development of atomic power reactors but on the willingness of private insurance companies to cover such plants. Further, Cisler had launched his own studies at the University of Michigan to determine what would happen if there were a major nuclear accident at the Fermi site itself. Preliminary rumblings about what both the AEC and the University of Michigan studies might come up with were not at all reassuring. Some rumors were that estimates for the worst possible type of accident at the Fermi plant would run to more than 100,000 killed.

It would take until the following year—1957—before the official figures on both studies would be released.

By the end of the third day, the Advisory Committee on Reactor Safeguards finally reached a decision. Rogers McCullough, the chairman, sat down on June 6, 1956, to carefully frame a letter to K. E. Fields, the general manager of the AEC in Washington. Fields was responsible for running the commission, under Chairman Strauss and the four other commissioners of the AEC board.

McCullough began the letter with the guarded statement that the proposed Fermi plant represented a greater step in the state of atomic power than any existing reactor. Then he listed the conclusions of the advisory group:

"1. Even though there are no facts or calculations available to the Committee that clearly indicate the proposed reactor is not safe for this site, the Committee believes there is insufficient information available at this time to give assurance that the PRDC [Fermi] reactor can be operated at this site without public hazard.

"2. It appears doubtful that sufficient experimental information will be available in time to give assurance of safe operation of this reactor unless the present fast [breeder] reactor program of the A.E.C. is amplified and accelerated as detailed below.

"3. It is impossible to say whether or not an accelerated program would give sufficient information to permit safe operation of this reactor at the Lagoona Beach site on the time schedule presently proposed."

McCullough then went on to list the steps that would have to be taken to provide enough information to make an informed judgment. The power surge problem with the EBR-I, for instance, would have to be clearly understood so that there would be no danger of a similar accident happening at Fermi. The EBR-I accident had been within a half-second of disaster. If anything like that happened in the Detroit area with the larger Fermi reactor, almost certainly the future of atomic power plants would be cut off before they could get off the ground.

McCullough also noted that provisions would have to be made to keep the fuel elements of the Fermi reactor from bowing out of shape, as had happened in the EBR-I. Any time the thin rods bearing the fuel moved into a more compact shape, the splitting of the atoms could intensify and begin to run away.

There would also have to be assurance that any start-up tests slated for the Fermi site would be checked out in advance, in an unpopulated area. Unless a "negative temperature coefficient"—needed to wipe out a sudden surge of power—was available to prevent a meltdown, no chances could be taken. The whole pre-testing program should be fully verified to make sure that there could be no meltdown "under any conceivable circumstances of control mal-operation," as McCullough put it in his letter.

He continued by commenting that the Advisory Committee as a whole "was not satisfied with the evidence presented that no credible supercriticality accident resulting from meltdown could breach the container. It is felt that a more extensive theoretical and experimental program to examine all the possibilities needs to be established and pursued vigorously. . . ."

Most important, McCullough pointed out, were "mock-up experiments, to insure subcritical distribution of melted fuel, and to assure that free fall of core parts cannot reassemble a critical mass suddenly." In other words, if the fuel melted and suddenly dripped into a thick mass at the bottom of the core, an explosion could result that would be unpredictable. (Some studies have noted that the fuel could become compacted—a deadly condition—as it bashed into the other metals in the core to cause one explosion, which in turn could cause another one much larger than the first. A small meltdown or small explosion would not necessarily spell the end of the danger.)

McCullough congratulated Cisler's group for its pioneering effort to advance nuclear power, but concluded: "The Committee does not feel that the steps to be taken should be so bold as to risk the health and safety of the public."

The letter boiled down to three things. First, the planned Fermi reactor would probably be a threat to public safety. Second, there wasn't enough experimental information to make it safe. Third, the situation could not be corrected in time for the schedule proposed by Cisler.

The moment it was received by Strauss, the letter was marked "Administratively Confidential." For three weeks, neither the public nor the Joint Committee on Atomic Energy heard anything further about whether the construction permit for the Fermi plant had been approved or turned down. Plans for the Fermi reactor seemed to be floating in limbo. No construction permit had been issued, but neither was there any word of rejection.

On June 28, twenty-two days later, Chairman Strauss went to testify before the House Appropriations Committee on an entirely different matter. During the routine questioning by the congressmen, Strauss casually mentioned that he was going to attend the ground-breaking ceremonies for the new Fermi plant at Lagoona Beach, Michigan, on August 8—only a little more than a month away.

The full effect of his comment was not felt until the next day, when Thomas E. Murray, a fellow-member of the Atomic Energy Commission board, went before the same House committee and told them in no uncertain terms that the designs that the Detroit Edison Company had submitted to the Advisory Committee on Reactor Safeguards were not satisfactory, and had been turned down by that committee over three weeks before. He went on to say that the fast breeder reactor was classified by experts as the most hazardous of all the reactors. Further, no construction permit had been issued by the AEC.

Beyond that, Commissioner Murray made it known that Chairman Strauss had marked the Advisory Committee report secret, overruling accepted practices. Obviously, there was a storm brewing among the five AEC commissioners themselves, with Strauss and Murray squaring off for battle.

When the news of what had happened hit the Joint Committee on Atomic Energy, Congressman Chet Holifield and Senator Clinton Anderson broke out in unison to condemn Lewis Strauss's cavalier action. As chairman of the Joint Committee, Anderson demanded that the AEC make the safety report public immediately, since it raised serious questions about the safety of the Fermi reactor, and that it was so close to Detroit. He said the Joint Committee would get at "the full facts involved in the precipitate action" Strauss had taken. Holifield joined him in the protest, charging that the safety warning by the Advisory Committee had been deliberately suppressed.

Governor Soapy Williams and the fiery Senator Pat McNamara of Michigan immediately joined the chorus. Williams wired the AEC demanding that it release the safety report for public scrutiny. "It is my constitutional duty to protect the people of Michigan," the governor's telegram stated.

The only response from Strauss was that there would be "no public answer" to the telegram. Rumors had it that Strauss had agreed with Cisler that the Fermi construction should begin immediately, in the hope that the technical problems could be solved before an operating permit would be issued. Meanwhile, Strauss rejected Governor Williams's request to make the safety report public.

Senator Anderson kept pressing. He sent a telegram to Governor Williams, telling him not to take "no" for an answer. "Is the State of Michigan going to be kept in the dark?" Anderson asked in the telegram. If so, how would Michigan know how to guide its actions in dealing with the safety hazards that were obviously looming in the Fermi reactor?

It wasn't until mid-July, more than a month after the Advisory Committee's letter had been written, that Cisler's group made any comment. A brief statement to the press said Senator Anderson's allegations were untrue, and that the Advisory Committee on Reactor Safeguards had simply requested more information on the design of the Fermi reactor and more experimental confirmation of some of the factors involving it.

On July 18, the AEC finally decided to acknowledge Governor Williams's request for more information. K. E. Fields assured Williams in a long telegram that the "health and safety of the public would be protected" in issuing a construction permit for the Fermi reactor. But, on the grounds that it was a preliminary and internal document, Fields refused to reveal what was in the safety report. He went on to say: "I believe that there is some misunderstanding as to the function of the Advisory Committee on Reactor Safeguards, and the report it has submitted.

"The committee is a committee of consultants established by the commission for the purpose of giving the commission technical advice on matters relating to reactor safety. The commission has received advice from the committee in a letter setting forth certain views as to technical matters involved in the safety of the reactor which the Power Reactor Development Company [the official name for Cisler's group to build the Fermi reactor] proposes to build at Monroe, Michigan.

"The advice received from the committee will be taken into consideration along with the advice and views of other commission technical staff members on safety and other pertinent matters in passing on the application of the company for a facilities license and construction permit under the licensing provisions of the law and the commissions regulations."

The telegram seemed to forecast approval of the Fermi construction license. The assumption was correct. On August 4, 1956, the AEC issued the construction permit. It was just four days before the planned ceremonies at Lagoona Beach.

Outcries were not long in coming. Even though the AEC assured critics that the plant would not be allowed to operate until safety was certain, the protests grew. Senator Anderson called the AEC approval a "star chamber" proceeding, and announced that the Joint Committee of Congress would begin an immediate investigation. He said that the question of safety must be answered before permission was given to build, rather than afterward. With between $40 and $50 million being invested in the plant, he said, there would be enormous pressures to let the plant operate, even though safety risks still existed.

Holifield was also incensed. He called for a repeal of the license and demanded that President Eisenhower step in and rescind the construction permit on the grounds of inadequate public safety. He added: "If I were Governor of Michigan, I would take legal steps to prevent the construction of a reactor that has not yet been declared free of hazard." He noted that the accident of the EBR-I in Idaho had come perilously close to being a disaster, and that it was many times smaller than the Fermi reactor, and completely isolated from large communities.

Both Anderson and Holifield agreed that they would ask the next Congress to rescind the power of the AEC to grant licenses to private utilities. In addition, they would write into the AEC law a provision that all safety factors must be brought openly before the public before any permission to build or operate would be granted.

But Governor Williams was silent. He seemed to be reassured by the long AEC telegram. And Cisler's organization hailed the issuing of the construction permit as confirmation of their safety precautions. "We are confident," Cisler told a reporter, "that the reactor presents no hazard whatever. We would not think of building or operating it if we were not sure of this."

Regardless of the storm of controversy, the ground-breaking ceremony at Lagoona Beach on August 8 was an historic occasion. As the breezes from Lake Erie teased the flags and the bunting, the speeches glowed with optimism. Strauss was on hand, as promised, and the only thing tempering the festivities were the thunderheads that were certain to intensify as the first shovelful of Michigan dirt signaled the beginning of the construction.

The biggest thunderclap that followed the ceremonies came from Walter Reuther, president of the United Auto Workers. Senator Anderson had placed a call to him after Commissioner Murray's disclosure that the safety committee's recommendations had been ignored, and suggested that Reuther's union should bring a suit against the Fermi project.

It didn't take Reuther long to act. By the end of August, he had made a public charge against both the AEC and the Power Reactor Development Company. He demanded a public hearing based on the fact that the AEC had issued the construction permit against the advice of its own safety committee. He also reminded the public that during a recent congressional hearing, Strauss had called the fast breeder the most dangerous of all reactors. Referring to the letter from the Advisory Committee, which had finally been made public, he noted that the safety report had said that it was uncertain whether a meltdown and fuel reassembly could cause an explosion that would breach the container. Then he added: "In everyday language, this means that the reactor might convert itself into a small-scale atomic bomb." Beyond that, Reuther claimed that building such a plant would endanger at least three million people in a thirty-mile radius around the plant.

A Cisler spokesman countered by saying that the Fermi project had only received a construction license, and that an operating permit would not be issued until the plant had been built and checked out.

Almost inevitably, the issue began to shift from public safety to politics. Both Anderson and Holifield had been pushing for a public atomic power reactor program to supplement that of private industry, on the theory that nuclear power was too complex and costly for private industry to handle alone. A bill on this line had recently been defeated in Congress.

Cisler, usually quiet and soft-spoken, reacted vigorously to Reuther's demands. He tied them to the pressure to give private industry a back seat. "We are headed down the road to a socialist state," he told the *Detroit Free Press.* He later added that opponents of private power were "prepared to use any subterfuge to keep atomic power development in the hands of the government." He announced that the safety issue was a subterfuge, that the key to the controversy was public versus private power. "I think they are hitting below the belt," he said.

But there were others who felt that the issue was one of safety, and safety alone. They believed that the political controversy was only a minor offshoot, and that the reactors should be guaranteed safe whether they were public or private. As Reuther explored the ways and means of bringing charges against the AEC to halt the Fermi project, Michigan Senator Pat McNamara, an outspoken critic of the project, rose in the Senate chambers to say:

> Mr. President, I have been keenly interested in the efforts of the Power Reactor Development Company to build a fast breeder reactor near Monroe, Michigan.
>
> As a Senator from the State of Michigan, it is my job to be interested in what happens in the State of Michigan, particularly when it involves the Federal Government.
>
> Shortly after PRDC made its initial moves to build the reactor, I began to get suspicious of the motives of PRDC and the methods it was using.
>
> As time went on, it became apparent that PRDC was receiving not only aid and comfort from the Atomic Energy Commission—but something approaching outright collusion—to steamroller through this project.
>
> First and foremost in my mind was the vastly important question of safety of the proposed reactor.
>
> Safety, of course, is important when one deals with atomic energy in any capacity.
>
> But why is safety of such paramount importance at the PRDC reactor?
>
> One has only to look at the map of Michigan, Mr. President.
>
> Lagoona Beach, the site of this reactor, is just outside Monroe, a city of well over 20,000 inhabitants.
>
> And Monroe, Mr. President, is only about 30 miles from Detroit, a city of over 2 million, and surrounded by populous suburbs.
>
> That is why safety is so important.
>
> Up to now, the A.E.C. has appeared to run roughshod over

> the safety question. And there are many questions to be answered.
>
> Were these questions raised solely by laymen—who know little or nothing about the complexities and technicalities of atomic reactors?
>
> No. They were raised by the A.E.C.'s own Advisory Committee on Reactor Safeguards.
>
> And its questions to date have never been answered.

Aiming his righteous indignation directly at Walker Cisler, he set down his objections in a stinging letter to the Fermi reactor chief:

> Dear Mr. Cisler:
>
> Let me make my own position clear at the outset. I reject the myth you have concocted that this is solely a fight between public and private power interests.
>
> I would be more than happy if this work in the peacetime uses of atomic energy could be carried by private enterprise.
>
> I also, under normal conditions, would be extremely happy that this work was to be done in Michigan, with Michigan companies playing a leading role.
>
> What I object to is the sanctimonious approach you are making to the public.
>
> You attempt to put across the notion that this work can be done by private enterprise without a dollar of the taxpayer's money being involved—you underplay the safety questions—and you charge that any interference or questioning of your scheme is "politics."

Cisler responded by moving doggedly ahead with the construction. At the end of the month, the construction crews began pumping tons of liquid cement into the rocky base at the site—a process called grouting. It would fill the fissures and create a solid foundation for the reactor. By October 1, the excavations for the foundations had begun. By the end of October, the contracts for the huge reactor vessel—weighing 350 tons, and three stories high—had been signed. It would be a giant stainless steel pot, like an elongated pressure cooker, with walls up to two inches thick. This in turn would be encased in another pot with a domed top. By December, the pouring of concrete for the foundations had begun. Slowly, the project—now estimated at $50 million—was beginning to take shape.

As 1957 began, the momentum of the elaborate project was gaining swiftly. But the first week in January brought other developments too: The public hearings requested by Walter Reuther began. His United Auto Workers were joined in the battle by two other unions, the International Union of Electrical Workers and the United Paper Workers of America—both AFL-CIO. It was the first public hearing ever held on the safety of a nuclear power plant. With the construction already in progress, the stakes were high. Walker Cisler wanted to get on with the job, unhindered by defending the reactor at hearings or lawsuits. He was *convinced* that he could make the reactor safe after it was completed, in spite of the reservations of the Advisory Committee report. "In my opinion, it is the most outstanding project in the world today," he told a reporter.

Reuther, in a letter to Cisler, pointed out that he had demanded the hearing because the UAW had 500,000 members living in the Detroit-Toledo area whose lives could be jeopardized if the Fermi plant were completed and went into operation. He urged that Cisler cancel the plans for the Lagoona Beach site, and build a prototype fast breeder reactor at a remote site away from a large population. "A similar but smaller reactor operated by the A.E.C. in Idaho went wild in November, 1955," he said, "and for six months was so radioactive it couldn't be taken down for repair. The permit issued by the A.E.C. admits there is uncertainty as to whether there is a credible condition of meltdown and reassembly of the fuel of the reactor which would result in an explosion that would breach this gas-tight building surrounding the reactor." He compared the situation with the construction of power reactors in Great Britain. They were building one reactor at Dounreay, on the farthest tip of Scotland. They had built another at Windscale, in the borderland Lake District, away from cities and where the population was scattered.

The hearings began on January 8, 1957. They were to drag out for over two years. At the opening session, the Fermi group presented six expert witnesses, with Hans Bethe among them. Bethe insisted, in spite of many expert reports to the contrary, that the plant could be built without undue hazards to the lives and health of the public. In a counterattack, Reuther quoted the AEC's own Advisory Committee's warning about the lack of safety, and emphasized that an unproven and experimental reactor had no business being placed between Detroit and Toledo. He asked that the construction permit be rescinded.

Not the least interesting of Reuther's contentions was that the flight pattern for instrument approaches to the Grosse Ile Naval Air Station passed directly over the Fermi site. At least thirty-six approaches a day were made by the Navy planes. While it might be unlikely for one of the planes to plow into the reactor buildings, the results of such an accident would be so catastrophic that the question was: Should this be risked at the expense of the population?

Cisler's group presented 322 pages of written testimony in defense of the Fermi plant. Because of the volume of material, the hearings were adjourned to permit the hearing examiner to absorb all the arguments. As the hearings dragged into March, Cisler reiterated his stand that the arguments on safety would have no bearing until the plant was ready for operation in late 1959. Reuther reiterated his stand that the lives of an enormous number of people were at stake, and that the AEC's own experts had charged that even testing at the Fermi site was unsafe.

The question was still dangling as to what were the potential catastrophe figures for a major accident in a commercial reactor in a heavily populated area. But in mid-March, just as the steel construction work was beginning to rise at Lagoona Beach, the first estimates of potential casualties from a reactor accident were officially released by the AEC. They came from the WASH-740, or Brookhaven report, which dealt with the problem of what would happen if the engineered safeguards of a reactor failed to function, and the deadly fission products were released into the atmosphere.

The results were so appalling that even the most devout believer of fission power was stunned and shocked.

FOUR

When the Joint Committee on Atomic Energy had nudged the AEC into making the WASH-740 study, it had been hoped that the safety picture would be encouraging enough to reassure the private insurance companies and bring them into the fold. The Fermi reactor and the other nuclear plants planned throughout the country were still without any assurance they could get adequate insurance coverage. For the potential claims of all people killed and injured by an accident there was still only a token $65 million, timidly offered as a ceiling by a pool of the entire insurance industry.

Thus the results of the WASH-740 study had been awaited anxiously for nearly a year. The study group had been set up to try to define the outside limits of damage that would result from a major nuclear accident. To calculate the results, the group took the case of an imaginary reactor, situated thirty miles from an imaginary major city, on a large body of water, and in a low population density area. This happened to be very similar to the Fermi site, but was not intended to represent any particular reactor.

The study was to be objective. There would be no deliberate attempt to make it either unduly pessimistic or optimistic. It would be forced to examine the worst possible case, but the AEC admitted: "This study does not set an upper limit for the potential damage; there is no known way at present to do this."

The conditions assumed that the radioactive fission products in the core would have twenty-four hours after an accident to partially decay before the containment vessel was breached and the fallout released. In the case chosen, fifty percent of these fission products would escape into the atmosphere.

Even though an exuberant AEC public information man once tried to soften the ugly potential of fallout by defining the radioactive poisons as "sunshine units," any fission product inhaled or absorbed by the skin is deadly. And since they can't be seen, heard, felt, smelled (except by a rat, oddly enough), or tasted, these poisons are insidious. Among the most deadly of the fission products are Cesium-137, Strontium-90, Iodine-131, and Plutonium-239. There are others: halogens, rare earths, and what are called noble gases, because they refuse to mix in with the common herd of the atomic family. Some decay rapidly, and remain lethal for only hours or days. Others, like plutonium, take up to 24,000 years to lose half their radioactive potency. Ten tons of plutonium could produce nearly 200,000 billion particles of dust. Each particle is capable of producing lung cancer. Normally plutonium would not be released from a light-water reactor, but if even a very small quantity should be, it could be harmful.

Strontium-90 masquerades as calcium, and dives into the human system straight for the bones. Iodine-131 pretends to act like normal iodine, and goes for the thyroid and salivary glands. The entire ghoulish family of fission products emit alpha, beta, or gamma rays that have little respect for the cells of the body.

The figures that emerged from the group's carefully calculated studies were not encouraging. If the assumed accident happened under what is known as a common nocturnal inversion condition, the lethal cloud of radioactive gases and particles would kill an estimated 3,400 people within 15 miles of the plant. Severe radiation sickness would fell another 43,000 people up to 44 miles away from the accident. Another 182,000 people up to 200 miles away from the source would be exposed to a dose that would double the chances of cancer. Property damage alone would amount to $7 billion—about 10 percent of the government receipts at the time—in 1957.

The problem of evacuation—the only real answer to a massive release of radioactivity—was even more discouraging. From a hypothetical accident like the one proposed, 66,000 people would have to be rapidly moved out of a 92-square-mile area, stretching to a point as far as 100 miles downwind from the damaged plant. For slower evacuation, 460,000 people would have to be moved out of their homes, up to 320 miles downwind from the accident.

Projecting these figures to a major accident at an atomic plant near New York City, the accident could affect homes as far away as Pittsburgh, Buffalo, Portland, Maine, or Richmond, Virginia. If the same thing happened in a nuclear plant near Chicago, the effects could be felt as far away as St. Louis or almost to Des Moines, or Louisville, Kentucky. A major Detroit accident could, under these conditions, affect Toronto, Buffalo, Pittsburgh, or Chicago.

The release of these WASH-740 figures in March, 1957, brought on another storm of controversy, casting a shadow on the steel skeleton rising on the edge of Lake Erie. In a letter accompanying the report, the AEC hastened to explain that the casualty and damage estimates were unrealistically high, because they were based on the worst possible combination of circumstances. Critics of the AEC pointed out that the figures represented lower damage from radiation than could be expected, and that genetic damage and the danger of cancer had been overlooked altogether in the report. It also, they said, failed to take into consideration the radiation scattered back after it had been deposited on the ground.

Cisler was now facing not only the threat of Walter Reuther's lawsuit, but the possibility that he would be unable to get sufficient insurance to put the Fermi plant into operation, even if it was cleared for an AEC operating license. He still had hopes for his own study, then in progress at the University of Michigan, which was considering the possible effects of an accidental release of fission products specifically from the Fermi plant. But it would not be finished for two or three months.

The Joint Committee on Atomic Energy was aware of the problem of persuading insurance companies to insure against such mammoth risks. The committee wanted to foster atomic power, in spite of its concern over the untested breeder reactor. Not only were the private companies unable to offer coverage to the power plants, but they were making it impossible for any home owner to be protected from any damage to his house as the result of a nuclear accident. Every home owner policy excluded this sort of damage. The same applied to automobile policies. No coverage would be provided if an automobile were so damaged.

As a result of this impasse and in spite of their concern about the safety of a breeder so near Detroit, Senator Clinton Anderson and Congressman Melvin Price were pushing hard to put through what would become the Price-Anderson Act, providing government insurance for nuclear energy plants up to $500 million. In addition, the utility companies would be required to get as much private insurance as they could in order to qualify for this protection—to date the $65 million offered by the private insurance pool.

An interesting feature of the Price-Anderson Act—loudly protested by the critics fighting fission power—was that none of the utilities or manufacturers of the reactors would be responsible for any of the damage to the public beyond that ceiling. With $7 billion estimated in property damages by WASH-740, added to untold billions in death and injury claims, those suffering from the results would have to make it on their own beyond the $560 million ceiling. Not only were the utilities protected from financial loss that might result from their own negligence, but the taxpayers would be paying for their insurance through the Price-Anderson Act.

Those who supported the atomic energy program argued that without this government-subsidized insurance, not a single reactor would be built by private industry. The counter-argument was that if the reactor builders and operators didn't have enough confidence in the safety of the atomic plants to assume responsibility for the risks, they shouldn't be building them at all.

Meanwhile, welded sections of the reactor building were swung into place, and the structures at Lagoona Beach began to take shape. At the end of May, 1957, 10,000 people swarmed over the Fermi site for an open-house to present the project to the public. Locally, there had been no opposition. In fact, an inter-county board of town supervisors from six surrounding counties passed a resolution favoring the Fermi plant and praising its objectives.

Less than two months later, in July of 1957, the University of Michigan issued its study on what would happen if the fission products were accidentally released from the Fermi reactor. Any hopes that this study would be more encouraging than the infamous WASH-740 report were shattered.

The report had been conducted by Dr. Henry J. Gomberg of the Engineering Research Institute of the University of Michigan at Ann Arbor. Gomberg was an articulate and soft-spoken nuclear engineer who looked like a successful businessman and spoke of the intricacies of the atom in calm, measured tones. He was director of the university's institute for the development of peaceful uses of atomic energy, as well as professor of nuclear engineering. The report was to become known by his name—the Gomberg Report.

In his attempt to evaluate the likely effects of fission products on the surrounding population, several possible conditions were assumed. Critics, however, could not help but focus on the most pessimistic of the situations studied when the report was finally circulated. This involved the release of all the poisonous fission products during a time of temperature inversion, where a warm layer of air would clamp the cooler air to the ground like a lid over a box.

If the poisonous plume of radiation moved steadily toward Detroit, traveling on a wind coming roughly from west southwest, the estimated number of people receiving 450 rads—the level at which half the exposed population would be expected to die—came to 133,000. (Rad is the nomenclature for Radiation Absorbed Dose.) Another 181,000 would receive 150 rads, with nausea and the probability of leukemia or other forms of cancer probably tripled within 10 years. Nearly 250,000 people would receive 25 rads—a dose that could be allowed in an emergency, yet represented an undesirable level.

In an inversion condition with a 4-mile-an-hour wind, it would take 20 hours for the plume of radiation to travel 80 miles. Any point on this circumference would sweep to the Indiana border on the west and Cleveland on the east. In Michigan, it would reach almost to Lansing on the west, Flint to the north, and Port Huron to the east, along with Sarnia, Ontario.

In spite of the pessimism of the report, Cisler and his team of engineers continued to move forward at full speed, confident they could avoid any fuel accident that would create these unspeakable conditions. The Gomberg-University of Michigan study was not publicly circulated at the time, and the WASH-740 report received little attention from the national press. The reports were extremely technical and filled with massive charts, formulae, and graphs undecipherable by the layman. Few critics of nuclear power were able to grasp the full significance of the potential dangers.

The rationale for going ahead with the Fermi reactor and the other nuclear projects across the country in the face of these awesome estimates of potential catastrophe was based on several reasons, which were often repeated and reviewed. Typical was a statement by Alton Donnell, who headed Cisler's design group, the Atomic Power Development Associates, at a Detroit press conference. "There is no question that the potential hazard due to the release of fission products from an atomic reactor can be great. However, reactors are designed such that, even in the event of a most extreme power excursion, no large energy release is involved. And multiple lines of containment are provided to prevent the release of fission products to the atmosphere. . . . Furthermore, the designs, in which all reasonable precautions have been taken, are reviewed by the Atomic Energy Commission, and actual construction is inspected by the Commission and other government agencies. Supervisors and operators are trained and checked before they are permitted to operate a reactor."

The position of the critics, typified by George Weil, the consultant who had advised his client against joining the Cisler consortium, was this: AEC assurances to the contrary, there was a clear-cut risk of a nuclear explosion in a fast breeder reactor, along with a very clear possibility of breaching the plant containment structure. "No matter how it is phrased," Weil stated, "*nuclear and explosive energy, rapid reassembly of the fuel into a supercritical configuration and a destructive nuclear excursion, rapid core meltdown followed by compaction into a supercritical mass, or compaction of the fuel into a more reactive configuration resulting in a disruptive energy release,* the meaning is clear: Liquid metal (sodium) fast breeder reactors are subject to *superprompt critical conditions.* And, as the A.E.C. well knows, this technical terminology translated into layman's language is an *atomic bomb.*"

Other opponents of nuclear power felt strongly that the billions that would be spent in fission power plants could be channeled into alternate sources of energy—sources that wouldn't offer such harrowing dangers that fission accident and the eternal burial of plutonium wastes would entail. For a stop gap, there was enough coal to last for nearly a thousand years. Meanwhile, the problems of solar energy and fusion reactors could be worked out. Coal liquification or gasification was slow, expensive, and awkward—but much cheaper than a nuclear accident.

Further, the critics continued to insist that there was no such thing as human infallibility, either among welders or AEC inspectors. Beyond that, they emphasized that all the engineering skill in the world could not prevent an inept or psychotic control room operator from plunging the reactor into a hopeless cataclysm. Added to that was the very real potential of sabotage. And the critics repeated another protest: Why should the taxpayer pay for the insurance of a utility company. And why should these plants be built if they were so dangerous and uninsurable? Why should Congress set an arbitrary limit of $560 million for victims of a disaster, with only the $65 million token insurance coverage from the pooled insurance industry?

The Price-Anderson insurance act was passed by Congress in September, 1957, and one more obstacle in Cisler's course was removed. It was still uncertain whether the UAW would file its protests in the courts, rather than in the AEC hearing rooms, where the case was required to be aired first. It was highly unlikely that the AEC would rule against itself in the hearings, which were droning on monotonously over the months. Therefore, it appeared almost certain that Walter Reuther would seek a court-enforced injunction, the only step that would have sufficient clout to stop the Fermi plant from going into operation.

Meanwhile, the tall, muscular, cigar-chewing Walter McCarthy, with a loping gait and a passionate desire to get on with things, was still a prime spark plug behind the Cisler dream. He did not have much patience with pessimism or critics. But neither was he insensitive to what had to be done to make the Fermi reactor an example of the best possible answer to the energy crisis. He recognized that there was an honest but passionate difference between the two schools of thought about the safety of the Fermi reactor. He felt that the insurance supplied by the Joint Committee was a reflection of the opinion of the people. He was convinced that most people thought that nuclear energy was good for the country, and therefore good for them. He felt that if some people got hurt from an accident, the rest were more than willing to pass the hat to pay for the damages, because of the urgent necessity for more power. As a firm believer that business is business, he was convinced that, because a great portion of the country's taxes came from business, it was only fitting that the government foot the bill for the necessary insurance to get nuclear power off the ground.

As far as the ominous figures reflected by WASH-740 and the University of Michigan studies were concerned, McCarthy believed that any such accidents would be impossible because of the containment precautions. He did, however, share the common fear about creating plutonium. It could be stolen; it could be separated; it could be used as a threat. But to do so would take considerable amounts of money, equipment, and scientific brains. There was no question in McCarthy's mind that plutonium would have to be guarded carefully, even if it meant setting up a special armed force to do so. He was firmly in favor of shooting anyone who tried to steal Uranium-235 or Plutonium-239. But his conclusions all added up to the fact that he would rather have the needed electrical power and work out the plutonium problem later on.

Thermonuclear—or fusion—power had been proposed as an answer to the energy deficit. McCarthy felt it was too big a question mark. Though there would be no nuclear waste or danger of a meltdown, there had been no breakthrough in fusion. McCarthy agreed that, if it were available, it would be foolish not to choose it over fission. There was no doubt in McCarthy's mind that when the country was faced with five hundred fast breeder reactor power plants as planned for the future, it would literally change the face of society. But we would have to adjust to this half a trillion dollars worth of ultimate investment, regardless of the guarding and safety problems.

McCarthy moved swiftly in his grasp of the massive details of the Fermi project. Working first for Cisler's design company in reactor physics, shielding, control, safety, and computer operations, he moved effortlessly into the construction operation, the Power Reactor Development Corporation. Here his responsibilities took on operations, licensing, and finance for the Fermi plant.

He had a passion for detail, and he handled it well. In preparing first for the construction permit and later for the operating license, he had to continue to consider every conceivable contingency.

In the light of the two devastating estimates of casualties from a possible accident, he dogged every step of the planning and construction. Particular care would have to be taken with the control rods.

The EBR-I reactor had been saved by the dropping away of the entire Uranium-238 blanket, just half a second before tragedy occurred. But, in order to reduce the size and complexity of the reactor vessel that held the core, the Fermi reactor design would not have that sort of emergency backup.

Both McCarthy and his colleague Amorosi were aware that the boron control rods they planned to use would have some disadvantages, but they were not considered safety problems. Control rods, when they were dropped down into the core shut down the reactor by "poison control." The boron in the control rods served to drink up the fast spraying neutrons, stopping them from colliding with the fuel, and cutting off the chain reaction. In other words, the control rods "poisoned" the fission process, and stopped it from proceeding, unless, of course, there was an accidental meltdown.

One problem in the use of boron rods was that they tended to capture many neutrons that could be used to leak into the raw uranium blanket and produce plutonium, and thus created less new plutonium fuel. For this reason, the rods rested high above the core when the reactor was running, so that they wouldn't absorb too many of the neutrons. This produced another disadvantage, in that there would be a delay after a scram signal had been pushed to stop the reactor.

However, both Amorosi and McCarthy felt that there was greater reliability in the new design because of a more simple control rod design, which would more than make up for the disadvantages.

Beyond that, the Fermi reactor design called for ten control rods. Eight of them would be safety rods for an emergency situation. Only two of them would be used for starting up and shutting down the reactor under normal operating conditions.

The main concern the scientists faced was the unusual conditions that could lead to "an explosive release of nuclear energy," one of the terms that physicist George Weil had referred to in plain language as an "atomic bomb." Scientists working on a nuclear power plant always carefully avoided the use of that term. In fact, constant public relations statements were issued by the AEC that a reactor simply couldn't explode like an atomic bomb. How these statements could be made in direct contradiction to many sober and reliable studies by qualified scientists remained a puzzle.

However, McCarthy and Amorosi were basically prudent men, and they did not ignore the potential dangers, regardless of whether the possible explosion was called a "disassembly leading to a high-energy release" or "an atomic explosion." McCarthy's analysis showed that a prediction for these destructive effects for some reactors could equal an explosion of "up to a thousand—or even more—pounds of TNT," although he did estimate that this force would not exceed the force of five hundred pounds for the Fermi plant.

He went on to say in his analysis for the Fermi license application: "Of course, the calculation of a severe accident in a real power reactor means dealing with complex geometry and irregularities in composition. The conditions and cause of the reaction are really not known precisely, and the course of the accident prior to the large nuclear burst is usually very complex. Nevertheless, idealized calculations provide some gauge of the severity of the accident."

Aware of their very heavy responsibilities, the Fermi engineers examined and put into the plans every possible design factor that would eliminate the dangers as the project moved ahead. They considered what would happen if a fuel element were dropped into a just-critical reactor. They checked the control rods and looked at every conceivable situation in which the reactivity could be increased within a reactor. They reexamined the possibility of water getting in contact with the syrupy sodium coolant, causing a violent fire and explosion. Most important was what would happen if somehow the sodium coolant were prevented from flowing through the fuel subassemblies. This would cause the fuel to melt like an ice-cream cone on a blistering August day.

All through McCarthy's painstaking license application analysis, there are phrases that reflect the frustrations of many scientists:

"There are few answers available to these questions . . ."

"These are problems that remain to be solved."

"Theoretical prediction remains to be accomplished."

"Computational errors of 5% or 20% may be discussed, knowing full well that the inherent errors involved in idealizing the general problem are much greater."

"Hence, the situation on the equation of state for uranium remains obscure."

It was difficult enough for McCarthy and the rest of Cisler's team to handle the overpowering technical problems they were facing, without having to defend themselves against Reuther's persistent attacks against the wisdom of permitting the Fermi reactor to operate when its construction was finished. To date, Reuther's attempts to try to cut off the construction permit seemed to be getting nowhere.

Meanwhile, the structures on the Lagoona Beach site were rising on the edge of Lake Erie, proud and gleaming. The smooth, white dome of the containment building was completed on September 21, 1957. Because it represented the final stalwart barrier between a nuclear runaway meltdown and the safety of the public, it had received the most lavish care and attention.

It was stoutly built. The lower portion was a vertical cylinder, with one-inch-thick walls. All the joints were double-butt welded, and every welded joint was x-rayed. The building was tested under internal pressure of forty pounds to the square inch, then subjected to a soap film bubble test to check for leaks. The results were verified by an AEC engineer.

Elaborate environmental and meteorological studies were also carried out by the Fermi reactor crew. Monroe County sprawled over 562 square miles of mostly flatland. There were some 3,000 farms in the county, 250 of them dairy farms with sizable herds. There was the corn, the soybeans, the wheat, the orchards. An escape of radioactive fallout could be disastrous not only to people, but to farm products. Radioactive iodine and the lingering Strontium-90 could be ingested by cows grazing on the contaminated pastures, and passed along to children and adults through the milk. The same could happen with fish caught and eaten. Wells could be contaminated with radioactive wastes.

The University of Michigan meteorological studies, a further supplement to the hazards report, concluded that there would be ample time to warn well-users in the event of a release of contaminated liquid, and to warn civic and industrial users of Lake Erie water who had intakes downcurrent from the Fermi plant site. At worst, the warnings would prevent use of contaminated water and farm products, should any escape of fission products spew out over the countryside.

But just sixteen days after the soap film was spread over the containment shell to check for leaks in the Fermi structure, another shock wave hit the nuclear fraternity. And even though it happened thousands of miles away from Lagoona Beach, Michigan, its impact would be felt there and at every other nuclear installation planned or under construction throughout the country and the world.

FIVE

The two towers of the Windscale atomic reactors, more than four hundred feet high, rise on the edge of the Irish Sea, near a village named Seascale. To the north along England's Cumberland coast is Whitehaven, seven miles away, just at the point where Solway Firth turns in to form a bay that helped shape the borderlands of Scotland and England. Here, history is found under every rock. To the west of the giant Windscale towers lies the rapturously beautiful Lake District, hailed by Wordsworth for its meadows, yew trees, mountains, and streams. "Through primrose tufts, in that sweet bower/The periwinkle trailed its wreaths/" he once wrote. "And 'tis my faith that every flower/ Enjoys the air it breathes."

It was uncertain, in October of 1957, that the flowers or the rich, verdant farmland along the coast and in the Lake District were totally enjoying all the air they breathed. The nearby port of Whitehaven, an industrial and coal mining city, poured a generous amount of bitter industrial smoke over the region, while the Windscale nuclear plant released an undetectable flow of controlled effluent, all of it under the allowable limits of radiation, however.

During these first days of October, life in the Lake District and along the Cumberland coast was moving along at its gentle pace, enlivened by the fresh employment the Windscale atomic energy installation brought with it. Some 3,000 jobs had been provided, a boon to the laborer and to a tidal influx of physicists and scientists who blended into the sparsely populated region effortlessly. They brought new blood into the pubs and country inns, so much so that the tiny Scawfell Inn at Seascale was able to support an extra cocktail lounge at its seaside perch.

Very little notable activity was taking place during that week. Workington and Oateshead had battled to a one-to-one tie on the soccer field before a crowd of nearly 8,000. A forty-six-year-old mother of eight children blamed four drinks for inspiring her to lift seven wine glasses, a toy car, three pairs of socks, and a child's vest from the counters at Woolworth's in Whitehaven. She stoutly claimed, however, that she had at least paid for some sugar and sweets. At the village dump at Cleator Moor, it was noted that the rats were growing almost as big as the cats.

It was of mild interest that Alsatian dogs had recently been stationed around the perimeters of the atomic plant, but this was understandable because plutonium was being manufactured there as part of Britain's weapons program. Fortunately, the dogs were trained to bark rather than bite.

Outside the chain-wire fence that surrounded the massive exhaust towers of Windscale Pile #1 and Pile #2, sheep quietly grazed on the bald, rolling hills that ended abruptly at the sea. John Bateman of Yottenfews Farm, adjoining the village of Carterbridge, near the gates of the Windscale atomic works, was busy with his dairy herd in which he took a great deal of pride.

The gaunt towers of Windscale were landmarks; they could be seen from miles away. Their square shape set them off from the usual round chimney. At the top, a large filter gallery almost seventy feet wide jutted out from the slender stem of the chimney. It was large enough to house a narrow-gauge railway track to facilitate the changing of the giant, fiberglass filters after they became fully contaminated with the radioactivity brewed in the giant reactor below.

The filters were important. They stood as sentinels against a sudden surge of fission products that might accidentally puff up from a mishap among the 1,500 channels of uranium cartridges that with silent, imperceptible power split the atoms to make plutonium.

Windscale Pile #1 was quite a different sort of reactor from the Fermi plant at Lagoona Beach. Its neutrons were moderated by giant blocks of graphite, rather than liquid sodium. But like all fission reactors, Windscale dealt with uranium, a fuel so temperamental that it allowed very little margin for error.

The uranium cartridges in the Windscale reactors were surrounded by the graphite which looks like and is quite similar to the lead in a lead pencil. It is nothing more than a very pure form of carbon. It serves to slow down the neutrons in the reactor so they have a better chance of hitting the nucleus of the other uranium atoms.

Very early in the morning of Monday, October 7, 1957, the huge blowers that kept the uranium cartridges cool in the reactor were shut down in order to begin a process called a Wigner release, involving the graphite blocks in the reactor.

Over a period of time, graphite will soak up a lot of extra energy like a blotter as it undergoes a constant battering by fast neutrons. Five years previously, the graphite in the Windscale reactor had suddenly released this stored-up energy in an uninvited burst of heat. Fortunately, the uranium had not caught fire. But definite safety measures were taken to stop such an unwelcome event in the future.

At regular intervals the reactor was shut down, and the

uranium and graphite were invited to heat up very gradually under controlled conditions. This was done simply by shutting down the air that cooled them. In this way, the graphite could slowly get rid of its stored-up energy, and everything could then continue with no chance of a sudden outburst. The condition was named after physicist Eugene Wigner, and was sometimes called "Wigner's Disease."

Curing this strange disease and fever could be a tricky business. It was like cooking a roast in a hot oven. If it is too hot, the fat could catch on fire. The device used in place of an oven thermometer was called a thermocouple, but its function was the same. When the reactor was shut down early that October morning, every precaution was taken to make sure that it was completely closed down. The thermocouples, so critical in the process, were thoroughly checked and the bad ones replaced.

The preparations were meticulous and painstaking. It took all night and through most of the day until the plant was ready to start the Wigner release. At about 7:30 on Monday evening, the signal was given to start the nuclear heating. This was carried on through the night until Tuesday morning when the heating from the uranium was stopped. But after an hour or so, the operators in the control room noticed that, instead of coasting along in the controlled heating condition, the "oven thermometers" showed that the graphite temperature was dropping. It should have been rising under its own momentum.

The physicist in charge was aware that in the past the process often had to be jogged with an extra heating, so the order was given to repeat the process. For some unexplained reason, he did not have an operating manual or detailed instructions on hand, but it would be a simple matter to handle any problem with the instrument readings he had carefully checked. The go-ahead was given shortly after eleven on Tuesday morning.

But very suddenly, a rapid rise in the temperatures of the uranium cartridges was noted. Immediately, the cadmium rods were shoved into the core to let the fuel cartridges cool. Somehow the "feel" of the reactor seemed different—much the way a car feels to the driver when something indefinable is wrong. The graphite temperature continued to rise. The instrument readings were confusing and conflicting.

The erratic conditions carried through until Wednesday night. When it became apparent that the situation was not going to correct itself, some of the ventilation was cut, just as a furnace is damped down. Some cooling air was provided to reduce the graphite temperatures.

It seemed to work. But at 5:40 on Thursday morning, the radiation meters near the top of the smokestack and the filters soared high on the needles. Then, just as suddenly, readings began to drop again.

By 8:10 A.M. Thursday, the radiation meters as well as the graphite temperature began to rise again. Another attempt to cool the reactor with forced air failed. All it did was raise the readings of the radiation meters high up on the stack. The fumes going up the flue were radioactive now, and the filter instruments showed it. Tom Tuohy, general manager of the Windscale works, and Kenneth Ross, the national operations director, were trying desperately to analyze the situation and to decide what to do about it. All the signs now pointed to a burst uranium fuel element and to a uranium fire. This could lead to an unpredictably serious situation.

Outside the health physics building, the readings rose to ten times the normal count. The situation was not yet critical, but it was ominous. Somehow, they would have to locate any burning fuel cartridges, force them out of the reactor, and keep the pile cool until the graphite and the rest of the uranium cooled down. Tom Hughes, the works manager under Tuohy, struggled with Ross's help to bring into action a cumbersome machine known as a scanner. When passed over the enormous face of the reactor, the scanner could sniff out where the hot spots were. But the scanner was overheated, and could not be budged. Just the previous day, it had been checked and was in perfect working order.

New air samples showed more high radiation readings. There appeared to be a critically bad burst. The diseased cartridges had to be located and quickly removed. A cumbersome hoist platform, a clumsy sort of an elevator that ran up the face of the reactor, was made ready. Hughes and Ron Gausden, the reactor manager, donned protective suits and masks, climbed onto it, and elevated it to the spot where one thermocouple showed a high temperature. With extreme care, they opened one of the inspection holes and peered in.

There was no question about it. The uranium was glowing cherry-red, with blue flames licking at the graphite surrounding it in the huge, three-story-high concrete block that held the hundreds of round fuel elements lying horizontally in it in separate channels. Between 100 and 200 channels seemed to be burning. Six other workers were brought in—all that could fit on the hoist platform. Then, working in quick shifts, a series of sweating, half-suffocating crews tried to punch the flaming rods of uranium out of the huge cylindrical drum with long steel rods.

But the uranium cartridges were bent, swollen, and wedged tightly. All that the crew was able to do was to disgorge the surrounding channels to make some kind of a fire break. Various shifts worked all through Thursday, with only partial success. Meanwhile, tanks of carbon dioxide were brought in, in an attempt to cool the flames. But the heat was too great. The fire continued.

The last resort would be to drown the reactor in water. It was impossible to predict when the other fuel elements in the core would burst into flame, loading the high chimney's filter with an impossible burden of fission products: radioactive iodine, strontium, and other gases. At least the particles, it was hoped, would be held back by the filters. But even if they held, the gases would escape.

The use of water was an agonizing decision. But if the fire kept up, the only possible answer would be the evacuation of the whole region—the farms, the villages of the Lake District, the towns and cities and ports on the shore of the Irish Sea.

Health physicists were everywhere. Film badges were checked and rechecked by a special process on the spot. Personal dosimeters—devices that clip on the pocket like fountain pens and that record body exposure to radiation—were supplied to everyone. Extra staff was called in to man the infirmary. The men who peered into the open plug holes received extra doses of radiation in the face and head, but because their film badges were not worn on the head, the extra doses they received were indeterminable. The managers of the plant canteens were ordered to put their food under cover.

Meanwhile, other health physicists were scurrying around the twisted roads of the region in panel trucks to check the radiation readings in a wide area around the Windscale reactor. The readings were rising, and they were beginning to cause great concern to the health physicists. However, practically no one in the neighboring villages and farms had any idea that anything was wrong. If they noticed the health physics trucks, it was assumed they were on a routine radiation check, a fairly common occurrence.

The worried officials at Windscale watched the barometer—the wind direction indicator—and the temperature with considerable concern. The weather condition would be critical in the case of any massive escape of the deadly fission products.

Throughout the day on Thursday the ground wind was light and blew the effluent from the stack away from the shore and out over the Irish Sea. But by night, it shifted. It pushed the plume coming from the stack down the coast, past the village of

Seascale, through Millom, and on to the sizable Lancashire community of Barrow-in-Furness. It was a city of over 60,000, with immense steel works, flour and pepper mills, and many acres of docks.

More serious than the shift in the wind was the nocturnal inversion. The warm air, a few hundred feet above the ground, trapped the radioactivity coming out of the huge stack, so that it spread along the meadows, the pastures, the trees, the homes, the cattle, the livestock. A survey van, moving to the cinder railroad track at Seascale, only a mile or so away from Windscale, checked the gamma rays and found them reading above scale.

The citizens there were unruffled; they still knew nothing about what was happening. Another van moved to the north, toward Gosforth, Calder Hall, and St. Bees Head. The reading along Calder Farm Road came to ten times the acceptable level for continuous lifetime breathing.

The health physics manager made some fast computations. Evacuation might become a necessity. No public warning had yet been issued, however, and it was only reasonable to try to hold off as long as possible. The health specialists were worried about three kinds of danger: gamma radiation to the whole body, inhalation of the fission products, or ingestion of them through contamination of the food crops or meat and milk products.

Usually internal doses are worse than external ones. The skin doesn't absorb radioactive poison as readily as the lungs and intestinal tract. The first readings indicated that the area of most concern would be from ingestion. But the concentrations were not yet high enough for the health physicists to take immediate action. Much would depend on the amount of Iodine-131 deposited on the pasture lands. When ingested by the cattle, this poison would go swiftly into their milk, which in turn would strike at the most vulnerable targets: infants and children. It was ironic that there was no established tolerance level for radioactive iodine in milk.

There was also the fear of the deadly Strontium-90. A series of long consultations among the medical and health-physics authorities began, as the men in the control room continued to fight the fire and rack their brains as to the best course of action.

It was becoming more and more apparent that the reactor had to be flooded with water. By now, it was nearly four days since the treatment for "Wigner's Disease" had begun. The use of water would mean the ruin of the multimillion-dollar reactor, without question. But the alternative was unthinkable. Literally, the life of Wordsworth's countryside, as well as its citizens, was at stake.

Drowning the reactor in water would be extremely hazardous. The water could possibly create enough hydrogen gas pressure to smash the chimney filters and release the full-scale poisonous fission products into the atmosphere. It was decided to hold off, but all would be made ready. Large fire hoses were moved into position.

Rumors had now become rife, but actual facts were sparse. Farmers, workers, and their families were bewildered and uninformed. Those who had a brother or father at the Windscale plant knew only that something was wrong. The workers had been warned, but told that there was no hazard to the public. A public information spokesman for Windscale made a brief and cryptic statement: "There was not a large amount of radioactivity released. The amount was not hazardous and, in fact, it was carried out to sea by the wind. There has been no injury to any person. There is no danger of the reactor's exploding."

In fact, the radioactivity was not being carried out to sea by the wind, nor had it ceased to be a concern, as the announcement implied. It was coming down the coast, over land. The only other official news that filtered out of the plant was that the reactor "was likely to be out of operation for some months because of an accident in which some of the uranium cartridges became red hot." As the announcement was being made, the sister reactor at

Windscale Pile #2 was shut down so that the workers there could come to the aid of their sweltering colleagues who were desperately trying to unload the blazing fuel.

The Windscale reactor fuel rods were loosely packed, and not as likely to explode as were those designed for a fast breeder, such as the Fermi reactor. This, at least, was one favorable aspect. But the Windscale reactor was estimated to hold some 150 million curies of radiation. If it were all released, the fallout would be just slightly less than that from the atomic bomb dropped over Hiroshima.

People in the village of Seascale seemed calm enough, even though the word had been bruited about that some men had been sent home from the plant. A reporter from the *Manchester Guardian* asked: "Isn't anyone worried about radioactivity here?"

"It's too late to worry about that now," he was told by a farmer, in a wry tone. "And anyway, they say the wind is blowing in the right direction." But then the villager thought a moment, and added: "I am told they found more dust than usual in Seascale this morning."

All through Thursday, the men in the control room continued to hold back the fatal decision to open up the hoses on the blazing fuel. It was a Hobson's Choice. The continuous fire in the reactor would eventually crack the defense of the filters whether water was used or not. Some protection might still remain, however, from some of the particulate matter, if not from the gases. The effect of the water on the cherry-red uranium was a deep imponderable. The reactor could "bump" with a milder hydrogen blast, or it could backfire violently.

Whether the water would be used or not, at 1:55 Friday morning, Kenneth Ross picked up the phone and called the chief constable of Cumberland at Penrith, some twenty miles away. He told him that an emergency standby should be set up. Hundreds of policemen were roused from their beds. It was the first official notice of the real trouble brewing.

The same warning went out to hundreds of men on the night shift at the chemical plants next to the reactor. Work was to stop. The workers were to assemble in the canteens. Construction work on the new atomic installation, Calder Hall, was stopped. No one was allowed to go outdoors.

Through the night and until dawn on Friday, the control room staff watched the instruments for any hopeful sign that the temperatures might be falling. But by seven o'clock, it became obvious they could hold off no longer.

Now the change of shift was due. The night shift, huddled indoors in the canteens, would have to be released, and the day shift put safely undercover, before the hoses were turned on. A little before nine o'clock, Friday morning, the shift change was completed. The men of the day shift were indoors, under shelter. The crew on the hoist, still sweltering and vainly trying to unload the fuel from the reactor, was sent out of the building, along with all the other control room crew. Only Tuohy, Ross, and Bill Crone, the fire chief, remained in the reactor building. As Ross later told reporter Chapman Pincher of the *Daily Express*: "I was never so frightened in my life."

There was an eighty-foot ladder to the concrete top of the reactor. Tuohy and Chief Crone hauled the fire hoses up to the top, sweltering and choking through their masks. At just before nine o'clock, the first hose was turned on, very lightly, very cautiously. The men jumped behind a heavy steel door, and waited for the "bump." It didn't occur. They increased the water pressure and waited again.

There was no hydrogen explosion. But there was steam. Live and hot, it flashed up the chimney. However, the fire crisis was over. The hoses were turned on to high pressure to continue all through the day. The pile was cold by noon Saturday.

If the fire crisis was over, the escape of radioactivity from the stack was not. The picture was confusing, because the radioactivity counts were both rising and falling in an unpredictable, spotty fashion. The long series of tests and meetings of the health physicists had continued through Friday and into Saturday.

The assumption was that the main risks would come from Iodine-131 and Strontium-90. It gradually emerged that the radioactive iodine had come through the giant chimney filters as the main culprit, although strontium leakage was not ruled out by any means. The tests completed by the middle of Saturday afternoon showed that the radioactive iodine had definitely been deposited on the pastures and foliage, and was a clear threat to infants and children in the area. Danger to the thyroid would be critical. Just how wide an area was involved was still uncertain.

By late Saturday evening, the medical group agreed that no chances could be taken, and that milk in the area would have to be immediately confiscated. The Milk Marketing Board for the county of Cumberland was notified, along with the police. The target was twelve dairy farms within a two-mile area of the Windscale plant.

John Bateman, at Yottenfews Farm, was roused out of bed by motorcycle police at 1:30 Sunday morning. He was told to keep his milk inside the cans until the scientists could come and check for contamination. One by one, the local farmers were awakened and given their orders.

It was revealed that the sample of the Friday milk supply ran to six times the permitted concentration (arbitrarily established at the time) of the radioactive iodine. This was small comfort to mothers with young children. The medical officers insisted that the external radiation was not enough to produce genetic damage. The farmers were assured that they would be compensated for any losses they suffered, but one farmer said plaintively: "We've never worried about radioactivity until now."

In spite of the reassurances, more farms were added to the banned list as the radiation vans continued to monitor the region. By Monday morning, October 14, the list of milk seizures jumped from 12 to 90 farms in the area. By afternoon, the list had grown to 150 farms under the ban. Milk sales in Carlisle, about 40 miles away, dropped 15 percent as housewives shunned it. Meanwhile, over 10,000 gallons of milk had piled up in the dairies. Imported milk was brought into the area for children and infants.

By Tuesday, the milk alarm had grown to enormous proportions. It stretched down the coast to Barrow-in-Furness, a hundred miles away by road, and to Millom. Together, they represented a population of 80,000. The Atomic Energy Authority would only say that the measured level of radioactivity in milk samples "taken on a gradually extending survey has not fallen off as rapidly as was anticipated."

The banned area now covered two hundred square miles. The sales of canned and powdered milk soared. In spite of it, the citizens of the area took the matter with typical British calm. "You never know what's going to happen next, do you?" said a cheery waitress in Barrow. "You have to be so careful with these radioactivities."

The swelling supplies of the contaminated milk would have to be shipped to the Milk Marketing Board's depot in Egremont. Here thousands of gallons would be unceremoniously dumped into "sea sewers," through which the milk would vanish into the Irish Sea, with the very clear probability of contaminating the aquatic life. With the whole stretch of coast under the milk ban, the problem of holding the milk before it could be shipped to Egremont became acute. A vaguely issued announcement said that the farmers could feed the milk to calves, but this was in direct opposition to a statement by the chief of the Windscale operation who said: "While this would be entirely harmless to adults and to pigs, I would hesitate to give it to young children or calves."

There were also widespread fears about the water supplies. But the Atomic Energy Authority assured the populace that there would be no danger from that source. The health physicists began a series of blood tests on the cattle to see how much radioactive iodine had been absorbed by the animals grazing on the contaminated meadows. New warnings went out regarding animals that were to be slaughtered, instructing anyone killing an animal to remove the thyroid gland.

Farmers grew increasingly impatient with the vague and confusing information supplied them by the Windscale authorities. They wondered why it was all right to drink the milk on the Thursday of the accident, but not on Monday. Why had the ban been extended down the coast so slowly? What would happen to cattle breeding? What about the property values of the land itself? Meanwhile, the men working at Windscale and Calder Hall received hot buckets of water and soap to scrub with before lunch. But they weren't told just how much contamination there was around the installations.

The miners in Whitehaven held a protest meeting to complain about the possible radioactive contamination of the mines through the ventilation ducts. New workers were brought up from Lancashire as "unexposed" workers to labor in the more contaminated areas. Radiation exposure is cumulative. Those who receive a more than normal dose must be kept away from any contamination until a long time has elapsed.

The tons of water poured into the fire were also loaded with radiation. It added to the exposure of workers in the immediate area of the reactor. An RAF helicopter was brought in to make tests 150 feet above the top of the chimney, photographing the filters down through the chimney opening as it did so. In faraway Devon, some three hundred miles to the south, some farmers who were unlucky enough to buy some West Cumberland cattle were ordered by the police to destroy them.

The confusion continued all through October as an official inquiry was conducted by the Atomic Energy Authority. For the most part, the public attitude settled down to: "They must know what they're doing. They'd tell us if anything were wrong." One worker at the plant said with a twinkle: "We're all radioactive here. What we don't know, won't hurt us."

But others were less charitable. A local official said: "If things are bad, we want to know. And if they aren't, we've a right to be told in words we can understand." Reporter Judith Hart interviewed a scientist at Windscale who had packed his wife and children off to the south of England when the fire broke out.

"We've known for years what was going on," he said. "The accident has brought it into the public eye, that's all. But radioactive waste has been coming out of that chimney and landing all over the countryside since the reactor started up. Not just radioactive iodine, but strontium and all that other stuff. We're all right—we think. But we don't know how our children will be later on. But it's no use worrying about it. We don't worry—we've just got to get used to living with it."

Reporter Hart got two different answers to one question: Can strontium and cesium and other radioactive substances get through the filters in addition to the radioactive iodine? The official answer was: "Only a gas like iodine can get through. No particles can get through at all. So there has never been any question of strontium or any other fission product in particulate form getting through."

But the chief medical officer of Windscale had a different story. He said: "The filters cannot be 100 percent, of course. They hold back big bits of dust. But particles of micron size can get through." A micron is 1/1000th of a millimeter. Fission products of this size are deadly. The radioactive iodine can be released both as a particle and as a gas. The problem, here and elsewhere, was that because radiation damage is so stealthy it was impossible to assess fully.

The official inquiry was intense but muted. It had all the earmarks of a coroner's inquest. Only a portion of the final report was made public. The investigating committee concluded that it was unlikely "in the highest degree" that anyone was harmed by the fallout, even though radioactive strontium finally was discovered in it, in addition to the iodine. It was noted that the strontium levels "in some pastures" indicated that a "watch" should be kept on the milk from these places.

No single individual was blamed or punished. In fact, the staff was praised for devotion to duty, which they obviously deserved. The accident happened because of a group error that revealed "certain weaknesses of organization." The report added: "Certain gaps in our scientific knowledge were revealed, and require early attention."

What was discovered in the inquiry was that even though the thermocouples and other instruments had been thoroughly checked, several were in error. This led to maneuvers which increased the extent of the disaster. The meters read lower than they should have. But there was human error, too. The combination was disastrous, especially in the newly developing and rapidly expanding field of nuclear reactors where there was no room for fallibility.

The milk ban was finally lifted several weeks after the accident. The farmers were compensated. The check for possible Strontium-90 deposits continued. But Windscale would be closed and sealed forever as an operating reactor, its multimillion-dollar investment written off. Even so, it could not be touched or examined for ten years, because it was so radioactively hot. Hundreds of workmen, and millions of dollars, would be required to dismantle it safely. It would have to be guarded forever. The tall, gaunt towers remained as a landmark, looking out over the Irish Sea and back toward the mountains and streams of Wordsworth's Lake District—a region they had almost raped and violated. As one British scientist put it: "The towers of Windscale remain as a monument to man's ignorance."

The accident was a grim warning to all who were in the

process of expanding nuclear power. The engineers and scientists at the Fermi plant at Lagoona Beach, along with everyone else in the nuclear field, studied the accident intently. As complex as the Windscale reactor was, it was not as complex—nor as potentially dangerous—as the Fermi breeder reactor rising on the banks of Lake Erie in Michigan.

Whatever the types of fission reactors being designed or built in the world, they all faced the problem of dealing with uranium or plutonium, or both. The distances that the fallout from an accident could be felt was revealed when it was announced at a scientific meeting in London during the spring of 1958 that "an unusually large amount of fission products appeared in the air over London during the twenty-four hours ending October 12, 1957"—the day when the Windscale accident reached its peak.

Windscale is nearly three hundred miles away from London.

SIX

For Cisler and his group the Windscale accident meant doubling their safety efforts for the Fermi reactor, as the construction moved slowly but steadily toward the day when the first tests would be made. By April of 1958, the huge reactor vessel itself had begun its 1,500-mile barge and railroad journey to arrive at Lagoona Beach by the first of May. At the same time, Cisler and the legal and technical staff of the Fermi project were preparing a position summary to present to the AEC about the inherent safety of the reactor. The presentation would call for the continuation of the construction program as an essential part of the nation's reactor development program.

Even as Cisler and his associates were preparing their case, another ominous blast rocked the industry. It took place at Chalk River, Canada—this time with a reactor known as the NRU, a sister to the NRX that had threatened that lovely community almost six years earlier.

Since early 1958, there had been a series of problems with the NRU fuel rods, which were clad in aluminum sheathing. Fission products had been building up, as well as contamination in the tank system of the reactor core. In addition, the instruments for detecting problems in the fuel were beginning to become unreliable. On Friday, May 23, 1958, after a week of steady operation, the power started to rise inexplicably and the reactor suddenly shut itself down by automatic controls. The crew in the control room were at a loss to know why, but they decided to try another start-up. Immediately, the automatic controls scrammed the action. Several alarm systems went off.

There was evidence of high radioactivity in the heavy water that cooled the reactor. What the faulty instruments didn't show was that there was violent damage to one of the fuel rods—a situation of potentially great danger. But three fuel rods showed that they were loaded with hot radioactivity. It was obvious that they had to be removed from the core.

Removing a fuel rod—not to be confused with a control rod, which contains no uranium—from any reactor is a fussy and precarious chore. The NRU reactor required a giant railroad-type crane that rolled on tracks on a platform over the top of the core. Mounted on the crane was a two-story-high, tube-like affair called the fuel removal flask. It was filled with heavy water, and looked like a thin smokestack on a ship. When properly positioned, a hollow metal snout would slide down from the flask and slip into a hole at the reactor top, like an enormous mosquito sinking its proboscis into a victim. The snout would then clamp hold of the top of the long fuel rod and pull it back into the narrow belly of the fuel removal flask, which was filled with heavy water as a coolant. There the fuel rod would be allowed to cool until some of its radioactivity decayed. Then the rod would be dumped into a storage bay not unlike an underground swimming pool filled with ordinary water.

It was a delicate operation. An exposed rod releases deadly radiation, and can burst into flames unless it is cooled by water. After hauling one of the hot rods away, the crane returned on its railroad tracks to pick up the second one, the rod known as J-18.

But this rod was swollen and warped, and couldn't slip up into the flask. A bigger entrance snout had to be installed. In doing so, it wasn't noticed that the heavy water in the huge tubular flask had drained out through a broken valve.

Even a small piece of irradiated uranium fuel is potentially deadly. A single irradiated fuel rod exposed to the air could release some 10,000 rads or more each hour. It takes only 450 rads to kill fifty percent of the people exposed to it, if they are without protective suits. Any container or cask used for moving an irradiated fuel rod around must always be kept filled with a liquid coolant. The coolant's loss means inevitable disaster, since no steel container can hold back radiation without it. The rays have no respect for mere metal.

Late on Wednesday evening, May 24, the big tube flask was positioned exactly over the hole where J-18, the badly damaged fuel rod, was resting inside the reactor. Very gingerly, the rod was raised partway up, and brought to rest still within the heavy shielding of the reactor vessel. At this point, the crew discovered that the heavy water had drained out of the tube. There was no time to lose.

Only more water could prevent disaster, but some of the most critical pumps had automatically locked themselves off because of the loss of water. The operator on the railroad crane took the only possible action. He tried to shove the rod back into the reactor. It jammed. Then he hit the button to extract the rod again, while other members of the crew in protective suits and masks rushed to bring emergency hoses to the deck on the top of the reactor where the crane sat.

The damaged fuel rod had now been without cooling water for nearly ten minutes. The snout of the crane finally picked up the burning fuel rod, and telescoped it back into the tube flask. A signal light flashed on the panel of the crane. It indicated that at least the fuel rod was up inside the tube. The operator hit the switch to move the crane along the tracks to let the hoses get at it.

By now, the fuel rod had been without cooling for twelve minutes.

There were several safety devices on the railroad crane that prevented it from moving unless the tubular flask was operating properly. Certain switches could not operate unless others were off. These were called electrical interlocks. They were built to prevent certain maneuvers of the crane that could cause danger. Under the present emergency situation, however, it was necessary to risk the dangers, but the electrical interlocks would permit no such thing.

As the crane operator punched the switch to move the crane, the drive motors immediately stopped because of the safety interlocks. Almost at the same moment, the radiation alarms went off loudly. A control valve on the tube opened when it shouldn't have. Men, sweating in spacesuits, rushed to close it. Another safety interlock prevented this. By now the meters showed the radiation rising from a hundred rads each hour, up the scale to several hundred, climbing toward the lethal 450-rad mark. The entire supervisory staff was called from their homes.

The crew, counting on their protective suits and masks, jumped the safety switch, the way an auto ignition is circumvented with wire. The clumsy crane began moving its precarious cargo toward the storage "swimming pool." It reached the point where the emergency cooling hose could be attached. Ordinary water was hosed into the enormous flask to try to cool the viciously hot uranium.

Because of the broken valve, the water streamed through the tube, all of it heavily contaminated, past the red-hot radioactive uranium rod, and flooded down on the crane platform. The poisoned water then cascaded down to the main floor and into the lower basement levels. The crew, hiding behind the giant crane to shield themselves from the radioactivity, watched the snout of the tube closely. As it passed over the repair pit which was sunk in the floor of the crane gallery, they were horrified to see a short piece of the now-blazing fuel rod drop out of the snout and into the open pit.

All but a skeleton crew was ordered out of the building. The operator stayed with the crane and moved it to the opening above the storage bay, so the highly contaminated water could pour into the "swimming pool" storage area below. As the water gushed down the shaft, the molten uranium in the pit continued blazing, filling the building with deadly fission products.

Outside the building, managers, draftsmen, accountants, engineers, and bookkeepers—all of whom had not been constantly and directly working around the reactor and building up cumulative doses of radiation the way the reactor crews did—met with the plant supervisors to volunteer service in the emergency. The radiation fields directly over the blazing pit now registered over 1,000 rads an hour—an unquestionably fatal dose for any measurable length of time, with or without protective suits and respirators.

The amateur office crew was suited up with masks. Each was provided with a bucket of sand. The job: to run into the building, up a long, precarious steel stairway, dash to the pit, and throw the bucket of sand on the burning molten uranium. A scout was sent ahead to scramble up the stairway, spot the exact location and condition of the fire, and report to the sand-bucket man.

Then they went in—bookkeepers, managers, and scientists—and they didn't mind admitting they were scared. The first one in was an accountant. He poured the sand quickly over the fiery, misshapen fuel-rod fragment, dashed back down the ladder-like stairway, and out into the fresh air again. In the brief moments he was in the building, in spite of the protective clothing and mask, he had absorbed his entire permissible radiation allowance for the year.

The others continued, one at a time, like a grotesque track relay team on an obstacle course, covered with plastic suits and snoods, monstrous-looking Canadian army combat masks, rubber gloves over cloth ones, and slippery plastic overshoes over rubbers. With over 1,000 rads coming up from the pit, they were ordered to keep line-of-sight observations of the fire to a minimum. Several monitors showed that the radiation was so "hot" that it sent the meters off the top of their 1,000-rad scale.

The fire was out within fifteen minutes, but the lethal radiation was everywhere. A courageous crane operator went back to drive the snout down into the shaft of the storage bay to stop the heavily contaminated water that was still gushing out of the tube. Another inserted a plug into the hole where fuel rod J-18 had once rested. The clean-up job began immediately, just before midnight.

Using a borescope, which is like a flexible periscope of a submarine, they looked into the debris of the reactor vessel. They found finely divided uranium powder, which seemed to have sintered—welded together—into cinders from the high temperatures created by the accident. There was evidence of an explosion, perhaps from a chemical reaction between uranium and water. The blazing fuel of J-18 that had spread so much contamination was found to be just one small scrap of a rod of uranium, only twelve inches long.

The clean-up job was prodigious. The first problem was to get the scrap of uranium and sand, still lethally hot from radiation but no longer burning, out of the pit. Teams of six men, working only sixty seconds at a time, ventured into the building. Here they worked with twenty-four-foot-long hoes and shovels to scrape the sand and uranium onto a skid and then cover it with more sand. A large semi-trailer truck was backed into the reactor building. A four-foot-thick wall of concrete blocks was placed between the space for the uranium and the driver's cab. The area around the truck was so radioactive that no one could get near it. The crane operators, working in two-minute shifts, had to lower the skid onto it by touch.

It took until 8 A.M. on May 25 to safely lower the twelve-inch-long fragment of uranium onto the trailer truck. Every road in the area was cleared of traffic and people. Slowly, the truck with its tiny load of uranium buried in sand moved the one-mile distance to the burial ground. Each speck was vacuumed up, then the road surface was either washed with a fire hose, or the exposed part of the road surface had to be physically dug up, removed, and buried.

Other staff and office workers were called in to remove all the remaining sand and debris in the repair pit. The health physicists did some fast computing and agreed that it would be safe to let the workers take up to five rads of exposure—the maximum allowable annual limit. They worked in ghostly shifts of one and a half minutes each, fully armored with clothes and masks. Again they used the clumsy, twenty-four-foot-long hoes, rakes, and shovels, dumping the debris into garbage cans radiating up to two hundred rads as soon as the remaining crumbs of the uranium and sand filled them.

As the clean-up job continued through Sunday night, closed circuit TV cameras were installed so that supervisors could keep watch over the clean-up crews, and the crews could study their work areas in advance before entering the lethal atmosphere. Practically all the doors in the reactor building were sealed off. Throughout the first week after the accident, the radiation readings were terribly high—some still over 1,000 rads each hour.

Special suction equipment was employed—a device called a Vacu-blast which has a nozzle that could be manipulated by long holders. It was painful work. The vacuums would often clog. To remove even a radioactive piece of paper from the vacuum required the use of long poles with adhesive-coated tips.

By the end of the first week, almost all the plant personnel had taken all the radioactive exposure they could afford. In their place, the Canadian government sent in nearly three hundred members of the Canadian armed forces. They vacuumed, then repeatedly wet-mopped the floors, covering the cleaned areas with polyethylene sheets of paper. The basement areas, ravaged by contaminated water, were continuously wet-mopped with damp rags. The scrubbed areas were given "swipe tests" with filter paper. The filter papers were then brought to the radiation monitors for checking. Steeplejacks were brought in to scrub every inch of the eight-story-high walls.

The materials were those that any housewife would use: hand mops, rags, water, and detergent. Every surface, everywhere, had to be decontaminated.

About three months after the accident, the decontamination battle had been won. There was still more mopping up to do, but the radiation had fallen off to reasonably safe levels. Because of the precautions taken, no injuries were reported, although effects of radiation injuries can remain dormant for up to fifty years.

But the accident was sobering. There were some 1,000 fuel rods in the reactor. They were made of the unenriched, natural Uranium-238. If the fuel had been that of a breeder—enriched Uranium-235 or Plutonium-239—the effects of the accident would have been catastrophic.

As the final clean-up was being done at Chalk River, the fuel rods for the Fermi reactor were being fabricated by the Sylvania-Corning Nuclear Corporation in Hicksville, Long Island. In contrast to the chunky, lower-grade rods of the Chalk River reactor, the Fermi fuel rods, sometimes called fuel pins, were literally as thin as Fourth of July sparklers, and four or five times as long. But they packed a much greater wallop. To load the core of the Fermi plant would take a million dollars worth of rods, made of the potent enriched Uranium-235, which would be packed tightly into the small but powerful core. The tight packing and the richer fuel would increase the potential danger of the reactor, because the tighter uranium is packed, the more hazardous it is. But Walker Cisler, acutely aware of the safety problems, was constantly taking greater pains to explain to the communities the care that was being taken to make the plant at Lagoona Beach a model of safety. Speaking one evening to the Monroe, Michigan, Business Men's Association at the local country club, he said that full precautionary measures would be taken to bar the possibility of any sort of explosion. He cited the control rod system which would automatically shut down the reactor if improperly operated, the "negative temperature coefficient" which would also automatically shut off the reactor if the temperature and reactivity started to go out of control.

"Through these measures and many others which I have not mentioned," he told the businessmen, "we are confident the reactor plant presents no hazard whatever. We would not think of building or operating it if we were not sure of this."

Asked about the furor in Washington concerning the hidden safety report of the Advisory Committee, and the UAW protests against the plant, Cisler said: "It is a little hard to understand this controversy. I think it comes about largely through lack of understanding of the vast amount of work that has been done and remains to be done before the plant goes into operation. In our minds," he concluded, "there are no safety questions that cannot be resolved before the plant starts up."

There were few who doubted Cisler's sincerity. He was a man dedicated to the social good, and his motivations were honest. But subjective value judgments were involved, and this was what made the issue so difficult to resolve among men of goodwill on both sides of the fence. Whose judgment was correct, and how could it be determined?

Cisler, McCarthy, Amorosi, and the rest of the management staff of the Fermi project were competent, conscientious, and responsible people. If they had been building an ordinary, coal-fired generating plant, no one would have contested them. A violent accident or explosion in a plant like that could be expensive and could injure a handful of people, but it could not affect an area the size of a state, or kill thousands of people. This was where the value-judgment process came in. This was why the fight was intensifying between the critics and the creators of the atomic power plants.

Behind both sides there appeared to be two of the most driving forces motivating men: fear and guilt. Many of the advocates of fission energy harbored strong guilt feelings about the hideous threat that the splitting of the atom had hung over mankind. Harnessing the atom for peace would assuage that guilt. The opponents of fission energy had strong fears about permitting future generations to face a perpetual threat that would be caused by the thousands of nuclear power plants planned for construction by the year 2000. Their theory was that even if men like Cisler, McCarthy, and Amorosi turned out to be infallible, the men who worked for them, the manufacturers who supplied them, the shippers who transported the fuel rods, the inspectors who checked for quality assurance, the AEC safety committees who monitored them, could not all be equally infallible. Human error was impossible to escape, and safeguards were as fallible as the men who designed them.

At the time of the NRU accident, nuclear mishaps in the United States were showing a disturbing trend. Nine serious transportation accidents involving nuclear materials were revealed in 1958 by the AEC. One of them took place at Hanford, Washington, where a tank trailer carrying 1,500 gallons of radioactive uranium overturned. Its brakes had failed on a hill. Traffic was shut off. Firemen rushed to the scene and hosed the roadway. The contaminated fluid was flushed into a ditch. Painstakingly, the dirt was dug up and hauled away to be buried. Another trailer truck carrying uranium gas also overturned in Bardstown, Kentucky, with some escape of gas. The AEC claimed that no one was injured in any of the accidents, but this did little to allay the opposition's concern about extending nuclear power.

The long, drawn out proceedings instigated by Walter Reuther and the AFL-CIO were finally coming to a lumbering head before the AEC. There was a lot at stake. Fermi construction estimates now looked as if they would soar to the $70 million mark, nearly twice that of the earliest estimate. The AEC was supplying more than $4 million worth of research and equipment. Reuther and the UAW were already preparing to take the case to the courts, once the obligatory AEC hearings were completed. But the AEC hearings did nothing, in fact, to slow down the construction on the Fermi site. All through 1958, the steel skeletons of the various buildings were fleshed out with concrete. Construction began on the non-nuclear steam generating station. An "Atomic Information Center" was built on the site to inform school groups, educators, civic leaders, and professional groups about the project. Nearly 1,000 persons a week flowed through it.

Contracts were signed for three special steam generators, at a cost of $1,375,000. These were designed with special steel tubes to transfer heat from the closed-circuit pipes bearing the molten sodium from the reactor, hot enough to vaporize the water surrounding the tubes, without being exposed to it. Meteorological tests continued to check the air diffusion character of the Lagoona Beach location. The huge steel intermediate heat exchanger shell, thirty-one feet long and weighing twenty-three tons, arrived by truck from its manufacturer in Dunkirk, New York. Progress reports were continually being filed with the AEC and the Joint Committee on Atomic Energy, who reviewed them and called for more.

At the beginning of May, 1959, the reactor vessel plug, a 288,000-pound stainless steel "cork" designed to seal the neck of the reactor vessel, arrived from Combustion Engineering by rail. It was symbolic in one way, because it arrived just a few weeks before the final decision by the AEC was sealed and delivered. On May 26, 1959, the AEC ruled to continue the construction permit. The decision came as no surprise. The question now was, would Reuther and the unions bring action in the courts where the decision would be independent?

If that prospect worried Cisler, he didn't show it. He continued to build up a staff of highly skilled scientists and engineers. In addition to Walter McCarthy, there was William Olson, an electrical engineer with long power plant experience at Detroit Edison. Wayne Jens, the assistant technical director, and Eldon Alexanderson, the reactor engineer, were also high on the Fermi team.

These men, and others with them, were given concentrated courses, both in the AEC's own installations and in nuclear reactors in England. Nothing was spared in developing their expertise.

But with all this meticulous planning, problems began to crop up even before the nuclear fuel was brought to the Lagoona Beach site. In August of 1959, a series of tests was begun with the liquid sodium coolant that would be so vital in the operation of the reactor.

Pure sodium is such a tricky substance; it must be handled with extreme care. It is never found free in nature, and it is a good thing it isn't. It's a killer compared to its mild and useful cousins found in everyday things, such as table salt (sodium chloride) and baking soda (sodium bicarbonate).

In its original form, pure sodium is a dry, silvery powder. To attain the liquid form needed for the breeder reactor, it must be kept over 210°F. The liquid won't boil until it reaches over 1600°. Inside the reactor, the waxy, lustrous, shiny syrup keeps the fuel rods from melting and heats up to 1000° to create the steam for the electric power. It drinks up the radioactivity like a sponge, and becomes highly irradiated.

No leaks can be tolerated, because the moment sodium

comes in contact with air or water, violent explosion and fire result.

The design of the Fermi reactor was developed with great respect for this chemical killer, for sodium has many advantages, too. It transfers heat beautifully, and would not need forced circulation if the pumps failed. Its high boiling point would allow lower pressure inside the pipes and vessel. But if it boiled inside the core or if its flow was blocked, there might be a runaway meltdown. If it leaked from the reactor after it had become radioactive, the resulting fire and explosion would be disastrous.

That is why the preliminary tests were so necessary before the fuel was installed at the Fermi plant. They began quietly enough in an abandoned gravel pit about twenty miles north of Lagoona Beach. But on August 24, 1959, a sodium explosion suddenly erupted to blast the residents of nearby Trenton and Riverview. Half a dozen people were hospitalized, and scores suffered lesser injuries. Many homes were damaged.

It was a serious accident, but even more sobering was the thought of the disaster that would have resulted had the sodium been radioactive. As it was, it was a setback for the Fermi reactor, which, even in its incomplete non-nuclear state, was revealing itself as a prized tiger—beautiful, sleek, powerful, and awesome, but very necessary to watch, cage, and contain.

Safety devices were checked and rechecked by the Fermi engineers; others were added to, and the whole safety system strengthened. Aside from the problem of another sodium explosion, there were other considerations. The prevention of a meltdown was of the highest priority. There was the awesome possibility of the "China Syndrome," where the molten uranium collects at the bottom of the reactor vessel and melts through the earth.

The bottom of a reactor is really similar in a way to a giant coffee pot. The reactor core is held up from the bottom by support grids, not unlike a coffee percolator, except that it is lower in the pot. Under the reactor core is an empty space called the plenum. In the core over a hundred subassemblies are packed vertically; long, square wrapper cans each holding 140 fuel pins tightly in place. The sodium rushes into the bottom of the "coffee pot" under pressure, and squirts up through subassembly nozzles to cool the blistering hot fuel pins to keep them from melting, and to take the heat away through pipes to make steam. The sodium is then cooled in a closed-circuit system, and pushed back through the core again. Even though sodium is a "coolant," the rods are never cool in conventional terms. They are always hot enough to transfer 1000° of heat to the sodium, which in turn transfers this high temperature by sealed pipes to a boiler, where the steam is made to turn the turbines.

In the remote chance that the sodium coolant was blocked off and the pins melted, the molten mass would collect in the bottom of the coffee pot, either to continue melting in the China Syndrome, or to create an explosion.

To prevent the fuel from forming into such a dangerous mass was a primary, absolute safety necessity. Consequently, Amorosi and his designers had already provided for a cone-shaped stainless steel pillar on the bottom of the vessel, like a blunt, inverted ice-cream cone. It was about a foot high. If the worst should happen, and the melted uranium dripped like candle wax down to the bottom, it was hoped that the cone would spread the stuff out thinly, like a pancake, rather than having it form into the thick mass that could lead it to a melt-through or explosion.

For further protection, they designed a metallic sheet made from zirconium that would be spread across the bottom of the reactor. Zirconium is extremely resistant to molten uranium, and would help protect against the China Syndrome.

But if the bottom was protected, why not the surface of the cone itself? Al Amorosi thought it over. As an added safety protection for the cone, he decided it might not be a bad idea to cover the cone with zirconium plates. It was being super-cautious to do so, but with the accidents that were happening around the world with reactors, it wouldn't hurt to be too careful. Amorosi wrote a memo about the idea, indicating that it would be easier to implement it than to have to justify *not* doing so, when the AEC's Advisory Committee came around for another look.

But because of an oversight, the change was not noted on the "as built" blueprints. No one knew at the time that an ominous error was being made.

SEVEN

Two months after the AEC had ruled in favor of the Fermi operation, the U.S. Court of Appeals reviewed the AFL-CIO suit spearheaded by Walter Reuther. The brief was filed on July 25, 1959, but because of delays in the court process, oral arguments were not heard in Washington until March 23, 1960—eight months after the suit had been filed.

The shocking outcome was delivered on June 10, 1960, when the Court of Appeals ruled that the construction permit for the Fermi plant was illegal. Building would have to stop within fifteen days.

The experts in Cisler's office were thunderstruck. If the ruling held, the economic repercussions would be awesome. Cisler replied by petitioning for a rehearing within a week. The AEC joined in the petition. But on July 25, 1960, one year after the court action had been brought by the unions, the Court of Appeals denied the petition. Construction was to stop. Millions of dollars would be frozen in economic limbo.

Neither Cisler nor the AEC were ready to give up, however. Almost immediately they announced that they would appeal the case to the U.S. Supreme Court. The move was backed by the Department of Justice, which supported the AEC. While the case was being appealed, construction continued. Life went on as usual at Lagoona Beach as more giant, heavy equipment arrived. The molten sodium was injected into the full system. The construction permit was renewed by the AEC for another year. Mrs. Enrico Fermi, widow of the world-famous physicist, came to inspect the plant named in her husband's honor. The remaining months of 1960 swept by, with the Fermi crew doggedly moving ahead in spite of the Supreme Court appeal, which hung over the project.

On January 3, 1961, some 1,700 miles to the west of the Lagoona Beach construction site, the three-man crew of a reactor known as the SL-1, in Idaho Falls, was well into its duties on the 4 P.M. to midnight shift. The reactor was designed to be plunked down in the middle of the arctic to bring light and heat to remote military bases, so it produced only two hundred kilowatts, enough for about one dozen homes. Because it was small, it was being serviced by only three men at night. There was Richard Legg, in his mid-twenties, an electrician's mate for the Navy who had completed eight months of training at a military nuclear power school. He had worked for more than a year on the reactor. There was the reactor operator, John Byrnes, also in his mid-twenties, and an Army specialist. He had nearly a year and a half's experience at his post. The third man on the shift was Richard McKinley. He was only twenty-two, and a trainee, fresh from another Army training program.

The three men had lots to do that clear, cold night. The reactor building, looking like a fat, tall corn silo, sat on the flat plains about forty miles from the town of Idaho Falls, the AEC's bedroom community where the families of the three men lived. Though the reactor itself was small, it was part of a huge AEC testing station—an area covering 892 square miles, almost as large as Rhode Island. In addition to the SL-1, there were sixteen other experimental reactors scattered throughout the vast sagebrush and desert complex. By 9 P.M., however, most of the employees in the other facilities would be gone, except for the night crew and the fire and security personnel.

The SL-1 crew was working a lonely shift. To the west, the Lost River range loomed like a vague, dark silhouette. Highway U.S. 20 skirted near the southern boundary of the reserve; Idaho 88 paralleled it nearly thirty miles to the north. Both sliced partly into the AEC complex, and few headlights could be seen along the straight, flat stretches of the roads.

For the past two months things had not been going well in the belly of SL-1. The cadmium control rods had been highly uncooperative, with a tendency to stick and jam. Considering the SL-1's highly enriched Uranium-235 fuel, this was not a situation to be taken lightly. A critical and super-critical condition could emerge within millionths of a second. Worse, on several occasions, steam had seeped into the control room without warning. There was evidence of crud gathering in the coolant water. There was also swelling and bowing of boron plates installed on the fuel elements as a "poison" to keep the atom-splitting from going into a runaway chain reaction, by drinking up the excess neutrons.

The guts of any reactor take a beating, both from irradiation and corrosion. As a result, the tendency of control rods to stick had to be watched very carefully. In fact, orders had been issued to all crews that they must "exercise" the safety and control rods regularly to make sure they would respond promptly to achieve either a routine shutdown or an emergency scram. The exercising consisted of raising and lowering the rods from different heights to make sure they were running free. But by two days before Christmas, on December 23, 1960, it had become obvious that the reactor would need considerable maintenance work and inspection, and it was shut down for the holiday week.

To shut down the reactor, the control rods had been pushed down snugly into the core to stop the chain reaction. Part of the routine job required disconnecting the control rods from the motors and gears that hauled them in and out of the reactor. Only five of the nine control rods were in use at the time, the others remained in the core. The important one was rod number 9, which alone could start up the reactor from its central position.

Previous crews had completed most of the inspection and maintenance work by the time Legg, Byrnes, and McKinley reported for duty on the afternoon of January 3, 1961. They inherited a relatively simple job, as noted in their Night Order Book: To reassemble the control rod drives and prepare the reactor for start-up.

At some time before nine o'clock that evening, the crew scribbled a laconic notation in the logbook: "Replacing plugs, thimbles, etc., to all control rods." As casual as the log entry was, the crew would have to be extremely careful of control rod number 9. To connect it to the machinery that moved it, they had to lift the rod four inches by hand. This meant standing on top of the reactor vessel and hauling the rod up very carefully, so that this distance would not be exceeded. Although there was an ample safety margin in inches, a sudden tug on a heavy, sticky, seven-foot rod could yank it too high. If this happened, the reactor could surge out of control in a fraction of a second. But the crew had done this job before, and were well trained for it.

To work on this routine, the three men were all in the reactor area, which was connected by a stairway to the sheet-metal building that housed the now-empty control room, repair shops, and offices. Normally, there would have been nearly sixty men working on the regular daytime shift. The nighttime duties were light and made the larger crews unnecessary. The Combustion Engineering Company handled the management of the reactor on contract to the government, and they were not required to be on hand beyond the normal working day. However, John Byrnes, a qualified chief operator, could call on the civilian managers at any time of day or night if the occasion demanded it.

The first sign of trouble came at exactly 9:01 P.M., when an automatic radiation alarm sounded at the AEC fire brigade stations and the security headquarters several miles away from the SL-1. Immediately, the alarm was broadcast from the security headquarters communications system over the AEC private radio network connecting all the Idaho installations and staff homes. At the same time, the personnel radiation monitor at the gate house of another facility, one mile away from the SL-1, also sounded the alarm.

Forty miles away from the site, in the town of Idaho falls, Ed Vallario, the health physicist supervisor, was in the process of putting his children to bed when the radio alert sounded. He grabbed his Scott respirator and protective coveralls, picked up his colleague Paul Duckworth, and raced along U.S. 20 to the west, toward the reactor. At the same time, the AEC fire brigade and security forces speeded toward the reactor from their headquarters eight miles away, pulling up at the site at 9:10.

They were greeted by silence. The buildings were intact; the lights were still on. There was no fire, no smoke. No one was visible, no one greeted them. Security patrolmen opened the gate of the wire fence, moved cautiously toward the big silo that housed the reactor, then on toward the building where the control room was. The firemen went on ahead of them. They were wearing protective suits including two pairs of coveralls taped tightly at wrists and ankles, overboots, masks, and radiation meters. So far, everything appeared normal.

They reached the SL-1 administration building, watching their meters carefully. There was still no sign whatever of the three-man crew. There was only a ghostly silence. The assistant fire brigade chief cautiously entered the building. His meter, which registered only up to 25 rads an hour, went off the scale. He retreated. Within moments, a health physicist from a neighboring reactor arrived at the scene. He and a fireman cautiously moved into the building, toward the control room. There was still not a sign of the crew. Their meters smacked up to 25 rads, and they also were forced to retreat.

Shortly after 9:30, two more physicists from one of the sixteen other experimental reactors arrived, one of them with a meter registering up to 500 rads an hour—a potentially lethal dose to the unprotected. The new crew moved into the administration building, toward the control room. As they approached it, their meter jumped to 200 rads. They rushed back out of the building, then held a conference. With the radiation levels registering so high, the probes would have to be rationed among many different rescue workers. No one worker could be allowed to expose himself to this high a reading more than once or twice, and then only for a matter of seconds, even with mask and suit protection. In such situations timing is all important. Very brief exposures to lethal radiation can be tolerated with protective devices, but they are still dangerous.

Another probe by a new crew was made. This time, they dashed up the stairs to the entrance of the reactor building itself. It was a shambles. Burned and twisted metal was strewn everywhere. They could see none of the crew that had been on duty. Not even bodies. The crew's meters hit the potentially lethal 500 rads. There was nothing to do but retreat again.

If the reading were a lethal 500 rads at the entrance, it was obvious that inside the building the radiation would be viciously higher—far above the killer threshold. A dose of several minutes to a crewman, unprotected by mask and suit, would hit him with acute radiation sickness within half a day.

He would feel nothing at the moment of exposure, but then would come the ominous symptoms that forecast almost certain death: nausea, vomiting, weakness, followed by apparent recovery for a few weeks. Meanwhile, his red and white blood cells and platelets would be dying. His blood would be suffocating, as the oxygen in it depleted. Then there would begin bleeding from the nose, gums, and intestines, leaving the victim open for infection. The hemorrhaging would kill him.

It was not a pretty picture for the rescuers to face. But there were men in there somewhere, and the job had to be done. Vallario and Duckworth arrived about 10:30, just after the AEC-Idaho operations officer broadcast a Class-I disaster. The decision to enter the reactor building itself would have to be made by Vallario as the ranking SL-1 health physics supervisor. He was quickly filled in on the situation, the enormous radiation readings, and the fact that no men or bodies had been seen. Neither Vallario nor Duckworth wasted any time. They grabbed their Scott-Pak masks and dashed into the building. They knew the risk they were taking, but they allowed themselves three minutes. They assigned three other rescuers to stand by.

They scrambled up the stairs to the reactor building, and looked in at the shambles. Then they entered. Their meters soared to double the lethal dose—1,000 rads. As they moved in past the threshold, they saw two of the three men lying to the side of what had once been the top of the reactor. One was still and lifeless. The other was moving.

They picked up the man who was still alive and put him on a stretcher. Their three-minute allotment was almost up. They carried the stretcher to the top of the stairs leading down to the control room, then rushed out of the building to summon the standby crew. Within seconds, the crew of five were back. Part of the team checked the second victim who was barely visible. He was dead. The others picked up the stretcher, ran and stumbled out of the building to a waiting panel truck. The radioactivity from the man on the stretcher, who now seemed more dead than alive, was intense. The truck spun out fast to meet an ambulance at a roadblock established where Fillmore Boulevard met U.S.

20, several miles away. The doctor, fully shielded, examined the victim. He was now dead, with his body continuing to give off lethal radiation. It was not safe to take the body anywhere but back to the SL-1 site. The ambulance returned there with its tragic burden.

Meanwhile, another team scrambled to the reactor building, into the 1,000-rad atmosphere. The second body was still on the reactor floor, as if blown aside by the twisted wreckage. The third was still nowhere to be seen. Time was running out.

Then they looked up to the ceiling, one story above the reactor floor. The third crew member was impaled there. Part of the reactor rod was through his groin and out his shoulder. He was obviously dead. The rescue team left the building.

A decontamination trailer arrived at the scene. The rescue crews who had entered the building were stripped of their clothes, cleaned and washed, and rushed to the dispensary for further decontamination. Up to 30 rads had leaked through their clothing—not enough to present immediate clinical symptoms.

Further attempts at recovering the two bodies in the reactor building were stopped for the night. It would only expose more workers to extreme radiation at a time when the entire situation needed slow and careful assessment. No one knew what the chances were for a secondary nuclear accident. The radiation levels, for the time being, were so lethal that extreme care would have to be taken. All workers were ordered back to a roadblock established on U.S. 20.

It was not until 6 A.M. the following morning that the decision was made to remove the first body from the ambulance which by now was badly contaminated by the radiation emitted by the corpse. With extreme care, the clothing was removed from the body by a team of five men, heavily gloved, suited, and masked. Some of the clothing stubbornly stuck to the skin and hair. The body was still emitting up to 400 rads. Only the frame of the victim's film badge detector remained. It was impossible to tell how much he had actually absorbed.

Carefully, they placed the body back in the ambulance. The corpse was covered completely with lead aprons in an attempt to reduce the radiation. Then the ambulance was taken across the broad, flat desert toward the Chemical Processing Plant—the only place where it could be completely shielded. Here the facilities for handling deadly, used fuel rods were such that thick concrete sealed-off areas for "hot" fuel processing were available. With the amount of radiation coming from the body, conventional burial was out of the question at the time.

At the processing plant, attempts were made to further decontaminate the corpse, but it was useless. The radiation count remained inordinately high. All decisions about burial would have to wait. The body was packed in water, alcohol, and ice, in the hope that some of the uranium would leach out. Meanwhile, careful plans were made to remove the second body from the floor of the reactor. (It was now obvious that the removal of the third body, impaled on the ceiling, would take long days of planning before it was even attempted.)

To get the second body out of the reactor building, the crews rehearsed the planned routine carefully. Because the radiation exposure load had to be spread over many people, jobs had to be broken down into several steps, each team accomplishing only part of the plan. It took until 7:30 P.M. the following evening for this to begin. The maximum permissible working time was set at one minute for any individual. There were two health physicists and two military men assigned. One of the physicists held a stopwatch at the entrance to the reactor operating floor. The other stood by in the control room, where the body was first to be taken.

The rest of the four-man team rushed into the reactor floor. One took the shoulders; the other the legs. Their one-minute limit expired when they were halfway down the stairs to the control room. They kept on going, placed the body on a blanket in the control room, and retreated. Another team dashed in. They took the four corners of the blanket, and moved swiftly out of the building with it to a waiting ambulance. The second body was also taken to the Chemical Processing Plant.

There was still the third body, plus the condition of the reactor to be coped with. It was impossible to climb onto the structural beam next to the body. The beam itself was both heavily contaminated and precarious. The readings went as high as twice the lethal dose—1,000 rads on both the beam and the body. A photographer sent in to photograph the position was permitted only thirty seconds to film the grisly scene. It was obvious that it would take many days working under these conditions to extricate the body. An entire relay of teams was set up to begin the task.

First, the outside door was opened to allow a large crane to be positioned just outside it. Then closed-circuit TV cameras were positioned inside the reactor building. A large net was prepared and fixed on the crane boom, underneath the body in the ceiling. The TV cameras failed to operate properly, and one of the teams had to be wasted in order to see that the net was in a proper position. Four other teams were assigned to climb through the outside door to the height of the ceiling and free the body, so that it would fall into the net. Another crew handled the crane outside the building. No team was permitted more than sixty seconds in the reactor building. Six days after the accident, at nearly five in the morning, the body was dropped into the net, and the recovery operations were completed. The teams could now turn their attention to assessing what had happened to cause the tragedy, and what condition the reactor was in.

Fortunately, the SL-1 was not a breeder. It was not as dangerous as the Fermi reactor. Its fuel was less compact, it was a fraction of the power of Fermi or other commercial reactors, and there was no danger of a sodium explosion. Nor was there much possibility of a secondary meltdown. Quick preliminary tests, done immediately after the accident, had indicated that the core was subcritical—that is, shut down with no further chain reaction, and unable to sustain one. Fortunately, the release of fission products to the atmosphere had been reasonably light. There had been no fire, no burning of uranium to release the pent-up clouds that, even with the small size of the core, could have been devastating to the surrounding areas.

Small as the core was, the energy released by the accident was found to be the result of 1,500,000,000,000,000,000 atoms splitting within a fraction of a second. Over the next months, crews probing the mystery continued to run in and out of the area for sixty-second intervals. The radiation was so "hot," and continued to be so, that longer shifts were not permitted. The new crews were told: "Look—we can't send anyone in to guide you. When we say 'Go!'—go in and do your assigned job. When we hit the bell, no matter where you are or what you are doing—come out!" One executive was assigned a precarious job. A welder's torch had set fire to some cloth bags packed with round, lead pellets, which were used to screen off radiation from the open reactor. The pellets had spewed over the reactor building floor like buckshot. His job was to shovel up what he could into buckets. He was allowed forty-five seconds. He ran into the building. The pellets were everywhere, spinning him as if he were on roller skates. Barely able to keep his footing, he shoveled desperately, filling only one bucket. It was so heavy he could barely drag it to the doorway.

It was clear that a nuclear excursion—the euphemistic term for a nuclear accident—had taken place. This was surmised by the detection of many radioactive isotopes found on the victims' belongings: a sample shaken out of the clothing; a radioactive cigarette lighter screw belonging to one of the victims; radioactive copper in a watch band of another; activated gold found in a wedding ring worn by the third. All revealed varied fission products such as Cobalt-58, Chromium-51, Yttrium-91, and gross fission products. Other tests showed that the energy was not large—but it was large enough to lift the reactor vessel—a three-story-high cylinder as wide as a smokestack—out of its hole, and smash it against the ceiling. It was speculated that the nuclear excursion created a giant "water hammer" which smacked against the shield at the top of the reactor and lifted it up with tremendous force, and that it all happened within two to four seconds. The excursion itself was apparently over within 1/500th of a second.

What had caused the accident was still only a guess. The prevailing theory was that one of the operators had lifted the central control rod—number 9. Perhaps it had stuck, and he yanked too hard. Perhaps he had tried to "exercise" it again. Perhaps he had been under emotional strain, and was not concentrating on the job. Whatever had happened in that lonely spot on January 3, 1961, would never be known. The significance was that, again, men were not infallible.

As one scientist, T. J. Thompson of MIT, speculated: "Perhaps the operator decided to exercise the rod without thought as to the consequences of the action. It is also possible, however, that an operator in anger, in a moment of careless fun, or in an act of deliberate sabotage raised the rod suddenly. But all these are sheer conjecture." Fortunately, the SL-1 system was not pressurized, as in the commercial plants, and the fission product after-heat was not enough to make the core molten which would have caused further melting or fire. Less than ten percent of the poisons were released from the building.

The official probe by the AEC dragged on for months. As the inquiry began, the first urgent problem was that of the burial of the victims. They were still resting in a radiation-proof vault of the Chemical Processing Plant on the AEC's reserve, packed in water, ice, and alcohol in the attempt to leach the uranium and fission products from them. But by January 23, twenty days after the accident, the radiation count had dropped enough to consider giving the bodies a decent burial. Even then, because the exposed hands and heads had received so much radiation, they had to be severed from the victims' bodies and buried with other radioactive waste.

While the investigators were picking over the shambled carcass of SL-1, the first shipment of enriched Uranium-235 arrived for the Fermi reactor at Lagoona Beach, on June 9, 1961. The fuel pins, designed to breed 106 kilograms of plutonium each year, were almost a yard long, and clad in zirconium. They were thin, less than the thickness of a lead pencil, and 140 of them were packed into a stainless steel cartridge to make up a subassembly. They were separated by egg-crate-like supports. That would stop them from bowing or warping—two very dangerous situations that could cause an unpredictable chain reaction. There were 105 subassemblies to be stacked vertically into the Fermi core. But for the moment, they would be stored in the Fermi vaults until the AEC cleared an operating license to permit the loading of the reactor.

The Supreme Court appeal was now pending and a decision was expected soon.

Walter Reuther continued to assail the AEC for permitting the Fermi construction to continue in the face of both the impending Supreme Court decision and the fatal accident at Idaho Falls. Now he released a study of forty atomic reactor accidents, many of them minor, but all potentially serious, and linked this with the SL-1 tragedy. "This study," he said, "plus the explosion at Idaho Falls, confirm the validity of the trade union opposition to the construction of the untested fast breeder reactor near Detroit.

"The Detroit plant, built in spite of an appeal by intervening unions, is 300 times larger than an experimental model EBR-I which exploded in Idaho in 1955." Then referring to the SL-1 accident, he said: "It is clear that thousands of people would have been overexposed to radiation if the SL-1 had been built in a populated area, just as the fast breeder reactor is being built in the first commercial size in the Toledo-Detroit metropolitan area."

Coincidentally, the AEC issued its first accident report on the SL-1 on June 11, 1961—the day before the U.S. Supreme Court was due to give its verdict on the Fermi-Lagoona Beach case. The upshot of the AEC investigation was not one of great assurance to the public. It said: "We cannot say with any certainty what initiated the SL-1 explosion, and it is possible that we may never know." The report pointed out that the condition of the reactor core and control system had deteriorated to such an extent that a prudent operator would never have allowed operation of the reactor without a thorough review. It revealed that portions of the reactor had bowed and warped, and that the sticking of the control rods had been an old and familiar problem.

The report discussed the rod-dropping tests several days before the accident, which showed that three of the five rods simply did not drop as they should have, to cut off the atom-splitting in case of an emergency. Why the reactor was kept going under these conditions was never explained. The ultimate cause of the accident was only a theoretical guess: Control rod number 9 must have been pulled out too far and too fast. The specter of human fallibility was still stalking the peaceful uses of atomic energy.

Even in its death throes, the SL-1 remained dangerous. It would take months and years to disassemble the machine, and its radioactive wreckage and grave would have to be guarded practically forever.

The day after the AEC report appeared, the Supreme Court

Following the disaster at the SL-1 nuclear power plant in which three men were killed (pages 104-115), the AEC experimented with a small research reactor to see if it would actually explode in case of an operational error. It was a light-water model (see diagram, right) with a nuclear core measuring only 15 inches square and 24 inches high. Below, the interior of the reactor.

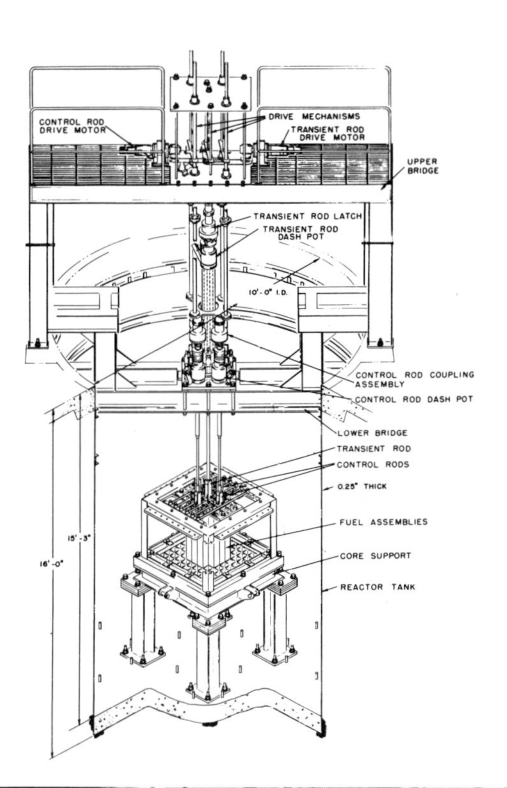

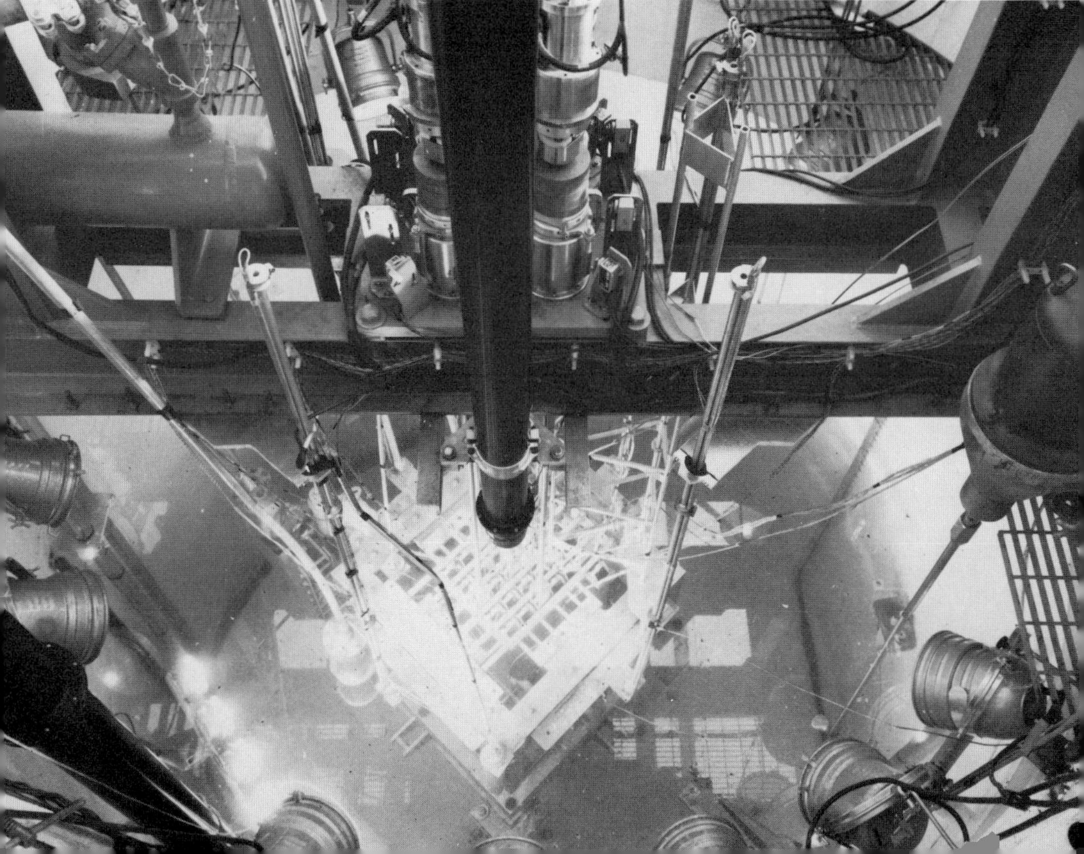

The tiny reactor, about the size of an orange crate, was blown up in November 1962. It zoomed from zero to 2.5 billion watts in less than one second and then blew apart in less than 1/3000 of a second, sending a cloud of water and debris 90 feet into the air. The small amount of radioactive contamination from this "clean" explosion was detectable 20 miles downwind. At right, the explosion; below, the mangled interior of the core after the experiment.

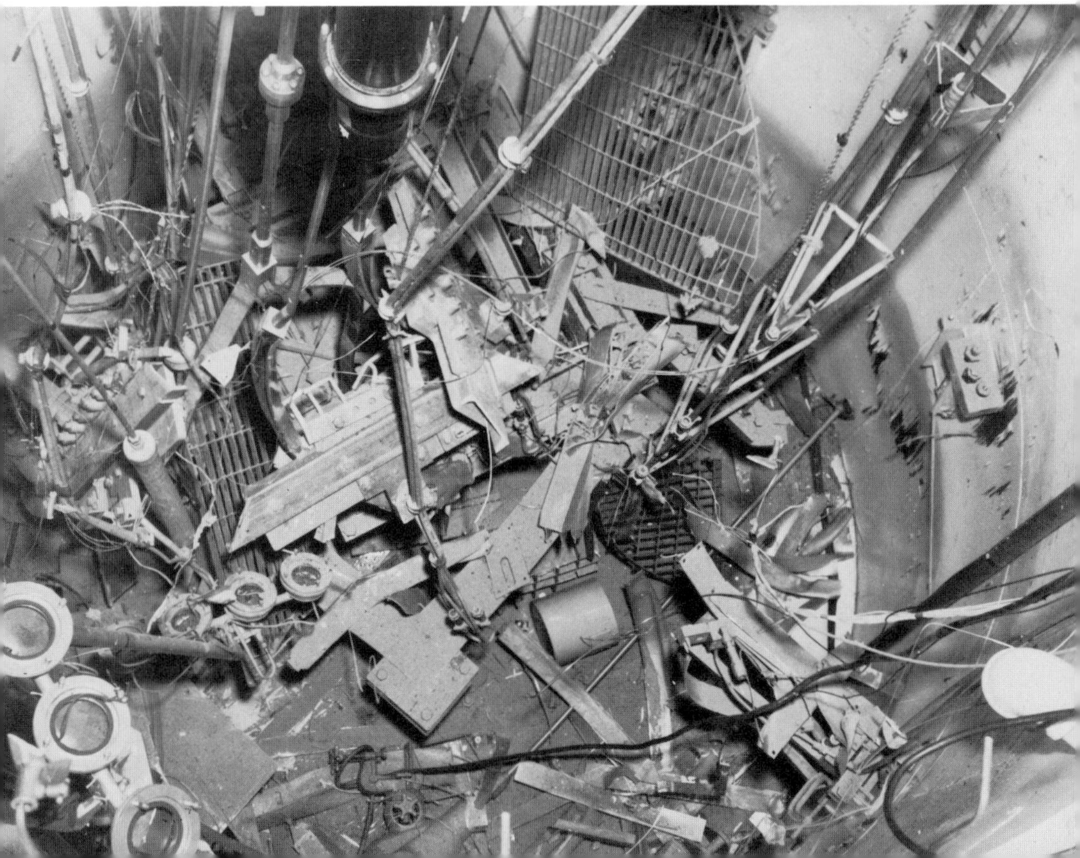

assembled in Washington to give its verdict on the Fermi case. Just what effect the disaster at Idaho Falls would have, neither Cisler nor Reuther knew. The case was to be the first contested licensing proceeding involving the AEC.

The struggle had begun in the summer of 1956. Now, five long years later, in the spring of 1961, it was coming to a head. It had cost time and money to both Cisler and Reuther—vast amounts. It would be a landmark case, the first that challenged the gargantuan capability of the atom to bring either enormous benefit or enormous catastrophe, depending on what was to happen in the future.

EIGHT

The Supreme Court decision came through on schedule. It was a clear-cut victory for Walker Cisler. The vote was seven to two. The decision stated plainly that the AEC had been within its rights in permitting the Fermi reactor to be built. Final construction could proceed unhindered.

In delivering the majority decision, Justice Brennan stated that the AEC had found "reasonable assurance for present purposes, and that is enough to satisfy the arguments of law." Brennan added that it was clear in the face of the law that Congress intended a step-by-step plan to provide the construction permit as the first step, to be followed by the operating license if the construction conditions were met.

Only Justices Black and Douglas dissented. They referred to the AEC's own safety committee report in their minority opinion which said: "Plainly these are not findings that the safety standards have been met. They presuppose . . . that safety findings can be made *after the construction is finished.* But when that point is reached, when millions have been invested, the momentum is on the side of the applicant, not on the side of the public.

The momentum is not only generated by the desire to salvage an investment. No agency wants to be the architect of a white elephant. . . ."

Then, referring to both the majority opinion that gave the Fermi plant a clear go-ahead, and the Atomic Energy Act, the minority opinion concluded: "This legislative history makes clear the time when the issue of safety must be resolved is before the Commission issues a construction permit." This decision, the report went on to say, "is, with all deference, a lighthearted approach to the most awesome, the most deadly, the most dangerous process that man has ever conceived."

The decision was a bitter disappointment for Walter Reuther. There would be some chance of challenging the operating license later, but the momentum seemed to weigh heavily against any possibility of stopping a project that had already reached an investment of some $80 million dollars.

The Monroe *Evening News*, elated that the project would continue to bring employment and additional taxes to Michigan and to the community, hailed the decision as "another notable stride forward." The local residents and press, however, had heard only one part of the story. The voices of protest were dim and distant. Hardly anyone outside of devout atomic-plant watchers had heard of Chalk River, Windscale, or the SL-1; few knew the implications of a runaway meltdown. Reuther's battle had been fought in Washington, far away from Lagoona Beach and Monroe, Michigan. the whole concept of nuclear power was too new, the dangers too obscure and technical for the lay press to grasp. Some newspapermen, like Saul Friedman of the *Detroit Free Press* did, but his was a voice in the wilderness.

The Supreme Court decision, of course, gave Walker Cisler and his team new life. A renovated buoyancy emerged at the site. Construction activity was brisk, and the rest of 1961 whisked by in a frenzy of activity.

As 1962 began, Walter McCarthy moved up to become the

assistant to General Manager Robert Hartwell, as well as serving as secretary and assistant treasurer of the Power Reactor Development Company, Cisler's combine. McCarthy's theoretical grasp of the intricate nuclear physics involved in the breeder matched his engineering and executive skills.

His preparation of the many stages of the Hazard Summary Report, required by the AEC before an operating license could be issued, was detailed and exacting. It was continually reviewed by the AEC. McCarthy was required to examine every possible type of danger.

His report exuded confidence and authority. "Even if a leak occurs in the primary system," he wrote, "cooling can be maintained and the core will not melt down. . . . Every precaution has been taken to prevent gross meltdown of the core with the possibility of an ensuing energy release."

Like most nuclear scientists in the field, he preferred to use "energy release" to "explosion." It seemed to be part of an unwritten code, just as "incident" was constantly used instead of "accident," "excursion" was used instead of "runaway," and "rapid critical assembly" was used in place of "potential atomic bomb." Having been faced with the slings and arrows of their critics, they were, of course, gun-shy and defensive. They could never tell when their words would be used against them.

In looking at what McCarthy considered the "maximum hypothetical accident," he was supremely confident that even what he called "this highly unlikely event" would not breach the containment vessel.

But other scientists, equally or even better qualified, were to disagree with this assumption some years later. George Weil was only one of many who was to equate a critical mass which created a blast with an atomic explosion. Professor Henry W. Kendall of MIT was later to reveal that the safety assurances put forth by the AEC for the light-water nuclear reactors were "gravely defective," and that the nuclear power plants being designed were a serious threat to the health and safety of the public. This applied, he was to point out in a later report, to all atomic power plants, without exception. His studies were to be resolved in a paper written in collaboration with Daniel Ford of Harvard, and issued for the Union of Concerned Scientists that was based in Cambridge, Massachusetts. They wrote:

> The safety systems in presently operating nuclear power plants are crude and untested. A number of design weaknesses in these safety systems have been confirmed. Moreover, there is extensive evidence that the workmanship going into nuclear power plant construction is far from adequate. The increasing number of quality assurance problems, maintenance deficiencies, management review oversights, and operator errors is disturbing. The A.E.C. has itself acknowledged that there have been a number of "near misses" in the brief operating history of commercial reactors, accidents that could have resulted in major public health incidents. An official A.E.C. assessment of some of the operating records of the nuclear reactor program is that absence of direct injury to the general public to date is "largely the result of good luck."

Rumblings against the breeder reactor also grew slowly and were to be singled out by prominent scientists across the country. Eventually, they were to group together to voice their protests. They included such prominent figures as Dr. James Watson of double-helix fame; Dr. Harold Urey of the University of California; Dr. Linus Pauling and Dr. Paul Erlich, both of Stanford; Dr. George Wald, the Nobel laureate of Harvard; Dr. John Gofman of the University of California at Berkeley, and many others.

Their combined statement against the fast breeder reactor was to be finally published several years later, and was uncompromising. "The reactor cooling system will utilize liquid sodium," they wrote, "which is highly reactive and burns on contact with air or water. Breeder reactors operate closer to the melting point of their structural materials, and they generate and use much larger quantities of plutonium. . . . Plutonium can be fashioned relatively easily into a crude nuclear weapon. In an energy economy based on breeder reactors (some hundreds of them by the year 2000 according to A.E.C. projections), enormous quantities of plutonium will have to be handled and transported. The potential for accidental release or theft by unauthorized persons will be unprecedented."

The statement added: "Federal funds being sought for the hasty demonstration and deployment of breeder reactors should be spent instead on such basic problems as reactor safety, waste storage and plutonium management. Of equal importance is increased Federal funding of other energy options, including solar power, controlled thermonuclear fusion, coal gasification, geothermal power, fuel cells, magnetohydrodynamics (MHD), and use of agricultural wastes and garbage. . . ."

The threat of future proliferation of plutonium did not shake Walker Cisler's confidence. He was looking forward to the day when the uranium in the core would be replaced by pure plutonium, which he considered more efficient in spite of the dangers. "That is what our goal is," he said to the Joint Committee on Atomic Energy of Congress. "We want to get to the point where we can fuel that reactor with plutonium. This is really what is behind our purpose. I am just hopeful that we can mobilize all of the know-how that exists anywhere in the world to enable us to put a plutonium loading in that reactor at the earliest possible time."

But as Saul Friedman pointed out in the *Detroit Free Press*, even Dr. Glenn Seaborg, the new AEC chairman and a co-discoverer of plutonium, called it the "ornery element." Its tricky chemistry, its capacity to create a flood of fission products at its birth, the need to process it in remote places, and its capacity to leap into a chain reaction, all were qualities that had to be weighed most soberly.

Even Dr. Edward Teller, one of the foremost nuclear proponents, was later to express his serious doubts about the breeder and plutonium as a fuel. "In order for a [fast breeder reactor] to work economically in a sufficiently big power producing unit," he wrote, "it probably needs quite a bit more than one ton of plutonium. I do not like the hazard involved." Dr. Teller was also to go on record as saying that not a single atomic power plant should be built above the ground, yet every plant built or planned was above ground.

Within a year after his promotion, Walter McCarthy himself, with David Okrent, wrote, in a section of the classic textbook *Technology of Nuclear Reactor Safety* (MIT Press): "At this stage [1963] of our knowledge of the course of violent disassemblies in large fast reactors of complex geometry, perhaps a word of caution should be added. The possibility that only a portion of such a reactor melts, undergoes a relatively mild explosion which acts to compress other parts of the core extremely rapidly, thus instigating a very much larger energy release, needs further investigation."

Even before the Supreme Court decision, and the euphoria it inspired, bad luck at Lagoona Beach seemed to be constant. Beginning in 1959, the fuel rod tests had shown that they would only be able to serve one-third of the hoped-for time before they burned out. The sodium coolant showed that it would strip the ribs that kept the dummy fuel pins at a safe distance from each other. The dummy fuel pins were for testing only. If they had been real, an alarmingly dangerous condition would have been created. In 1960, there was a four- to six-month delay, as tests showed that the fuel pins would swell and block the essential coolant from passing through the reactor. The potential power of the reactor had to be cut in half because of the tests on the fuel pin behavior. The sodium reacted with the graphite shielding, and much of the latter had to be replaced. It took fifteen months and $2.5 million to do so. The machinery dome design had to be changed, because it was found that the 288,000-pound plug to seal the top of the reactor could become a deadly missile and shatter the containment. All of this was reported to the AEC. There was no cover-up. The incidents simply dramatized the incredible problems encountered in this uncharted sea of complexity.

To add to these problems, in the fall of 1962 a subassembly stuck, more sodium plugged, more graphite deteriorated when it shouldn't have, and the enormous fuel lifting device failed. It took months to make the repairs.

The construction permit had to be extended, and the delays seemed endless, both technical and administrative. After all the years of monitoring the construction, the special Advisory Committee on Reactor Safety—the committee that had had so many reservations about the safety of the Fermi reactor, met in October of 1962 to give consideration as to whether the plant was now safe enough to operate.

Under the committee's scrutiny at the meeting was whether or not to load the fuel into the reactor, and begin tests that would use only 1/400th of its ultimate power. Because the reactor was going to operate at only a fraction of its designed power, the Advisory Committee cleared the way for permission to be granted to operate at this low level, subject to a thorough review before the power could be advanced up to its operating level of 200,000 kilowatts, the next step in its evolution.

But Reuther and the AFL-CIO unions had not given up. After having lost the battle against the construction license, they planned to fight the pending operating license at public hearings scheduled to begin in December of 1962. On December 11, they requested a delay in the hearing until January 3, 1963, to prepare their case. And, as if to punctuate the request, a dramatic event happened the next day—violating the law that had governed all the planning of the Fermi reactor: Sodium and air must never meet.

An operator sat at the console in the control room watching the instruments as the liquid sodium was rushing through the loop of massive piping that would eventually create the steam for the generator. Suddenly, the temperature rose on one of the dials. He reached over and hit a safety button to dump water out of the generator system. Only seconds later, an automatic safety disc burst. The sodium rushed out of a faulty relief vent. The moment it hit the air, it flared up violently. Fortunately, the nuclear fuel had not yet been loaded into the reactor, and the sodium was not radioactive. No one was hurt. But the unions protested vigorously that if the fuel *had* been loaded in the reactor, there would have been a disastrous release of fission products.

The accident brought the Michigan Attorney General's office onto the scene, and the public safety issue was dramatically thrust into the picture again. The sodium explosion was only a minor symbol of what could happen. The attorney general's office, along with other state agencies, were chary and circumspect about the whole matter. A provisional approval by the Michigan Department of Health was sent to the AEC, but it made clear that the state reserved the right to alter its position. The agency also indicated that the whole situation might have to be reviewed in the light of the sudden sodium leak.

The health department also asked the AEC to make sure that the monitoring devices for radiation be kept to strict standards, and that no leakage or radioactive materials be allowed to get out of hand. The department was especially interested in the emergency evacuation plans, not only because of the dangers of escaping radiation, but because of the contagion and health problems that could arise from moving thousands of people out of their homes.

The Radiation Emergency Procedures had already been

worked out by the Fermi plant staff, and they were elaborate. The shift supervisor would notify all plant personnel over the plant intercom, or by the blast of an air horn. The air horn signals were defined and worked out. Reactor engineers, operating staff, and health physicists would report to the control room to help the shift supervisor. Others of the staff would go to assigned shelter areas. Staff having to evacuate any building were to bring with them all radiation monitoring devices; plant guards would prohibit any but emergency personnel from going into the reactor site. Evacuation routes away from the plant would use both lanes, and traffic would be stopped to be monitored or decontaminated, at the Pointe Aux Peaux Road buildings, where first aid and sodium burn kits were available. The public information staff would be responsible for any plant visitors, whether they were at the Exhibits Building or on tour. All visitors would be checked for contamination. They would have their own film badges and pencil dosimeters.

Plant nurses and doctors were listed, and ambulance phone numbers posted. Any victim receiving more than twenty-five rads, or with a contaminated burn, would be taken by ambulance to the University Hospital at Ann Arbor, about twenty miles away. Specific officials—such as Sheriff Bud Harrington and the state police—were to be notified. They would be responsible for whatever measures were to be taken outside the plant.

The sodium accident elicited the outcries of the critics all over again. Several Washington reporters, including Esther Von Waggoner Tufty, protested that no approval should be given for the start-up of the plant until the explosion was fully explained. But the average citizen had only a little inkling of the potential dangers. In fact, the town fathers of Frenchtown, the local township just northwest of Lagoona Beach, wrote a comforting note to the AEC, saying that they backed the Fermi reactor a hundred percent. While some critics looked at the townsmen as mice walking into a bobcat's cage, the sentiments were genuine. "All questions have been answered," the town fathers wrote the AEC, "and the complete information which has been furnished has astounded us beyond measure. While not experts in matters of engineering and safety, we are impressed by the drive for excellence in these fields. Recognizing that safety, undoubtedly, is the most important factor in your consideration of these matters, we wish to pinpoint our favorable impression of the actions taken in the regard by the Power Reactor Development Company."

In spite of the sodium accident, the new year of 1963 began brightly for the Fermi plant. The AFL-CIO representatives were rebuffed at the new AEC hearings in Washington and the local township meetings concerning the operating license. In fact, the union men were so indignant that they walked out of the hearings, leaving the road practically clear for the Fermi license to be granted uncontested. But it still took five months longer for the license to become effective.

After a final AEC inspection, the Fermi crew began loading the Uranium-235 fuel assemblies on July 13, 1963. It was an eventful day, and the procedure had been carefully rehearsed beforehand. A cask car, running on tracks, and looking like a stunted white freight locomotive, was loaded with several subassemblies at a time. The operator, sitting in an open cab, manipulated a control board that moved the car and actuated a gripper encased in a housing. The car crawled along the tracks until it snuggled next to the huge shielded dome of the reactor vessel. Here the gripper would seal itself to an exit pipe. Then the cask car would slowly lower its expensive fuel assembly down the long pipe in the reactor vessel to a lazy susan sort of container, where the subassembly would be picked up by the offset handling mechanism—a giant automated lobster-like claw that would swing the fuel over to the round honeycombed grid that made up the guts of the reactor core.

Loading a reactor is like taking a bath in a pool full of crocodiles. A fuel loading accident could be as catastrophic in some ways as a reactor meltdown. If a fuel subassembly dropped, or was bent or damaged, there could be a disaster. The NRU accident at Chalk River, Canada, had shown what even a fraction of one fuel rod could do. It took six weeks to insert ninety-nine fuel subassemblies to prepare the Fermi reactor for the initial start-up. By noon on August 23, 1963, the engineers and operators and executives gathered in the control room. And as the Fermi plant was on the threshold of going critical, the pressures were building for a new study—one that would have a major impact on the entire atomic power plant front, including Fermi.

There was excitement and anticipation as the control rods began to be raised, cautiously, slowly. At exactly 12:35 P.M. the instruments showed that the chain reaction had begun—and could be sustained.

With the achieving of criticality, another long period of testing would begin before the reactor would be cleared for full operation. There were mountains of minutiae to be checked. The characteristics of the core would have to be plotted carefully in order to observe the behavior of the fuel under operating conditions. In the light of past experience, many bugs would have to be ironed out.

The problems continued through the first five months of 1964 in endless procession, with the reactor power still kept down to a fraction of its capacity. The cask car acted up now and then. The number 4 safety rod delatched, putting it out of action. A rotating plug stuck, making fuel changes impossible. A sodium pump was shut down for repairs. A new machinery dome had to be built, because the original didn't fit exactly. There were instrument problems, and sudden, dangerous, and unexpected gains in reactivity.

As the Fermi plant was undergoing its struggles and successes on the shores of Lake Erie, other, less complicated light-water atomic reactors were beginning to spring up throughout the country. In the year that Fermi was ready to "go critical," there were ten of these plants, with many others in the planning stage. They were of two types: boiling-water and pressurized-water reactors. They used lower grade uranium compared to the Fermi design; they were easier to build, and used water rather than sodium to cool the fiery uranium fuel rods.

Nuclear plants already were operating in Pennsylvania, Illinois, New York, Nebraska, South Carolina, Ohio, Minnesota, and Washington—in addition to Big Rock, Michigan. Most of these were not designed to produce more than 200,000 kilowatts of electricity, but they were necessary forerunners of future reactors. Although much larger than the small experimental reactors at Idaho Falls, the new designs eventually would have to produce a million kilowatts or more to be economical for the utility companies that would be building them.

Increasing the size of these reactors meant, of course, increasing the danger. In turn this would demand more and better engineering safeguards. It also meant that the AEC and the insurance companies would have to take a cold, hard look at the original WASH-740 safety report, which had been prepared back in 1957. For with new and larger atomic power plants looming on the horizon, the old estimates of 3,400 dead and $7 billion worth of property damage simply would not apply if the same type of accident happened to the newer giants.

On the other hand, many improvements in reactor design and safeguards had been made since the 1957 report. Perhaps these new safeguards could present a brighter safety picture for the public. In any case, it was obvious that the WASH-740 study would have to be brought up to date. Only in this way could a realistic picture for future insurance legislation be obtained.

The Joint Committee on Atomic Energy in Washington realized the importance of this, and in the spring of 1964 it instructed the AEC to set up a new study. The Price-Anderson insurance act was up for reappraisal, and it would have to be extended in some form, or the entire atomic power program would grind to a halt.

The responsibility for the new study fell on the experienced shoulders of Clifford Beck, the deputy director of regulation for the AEC. While he would head a steering committee for the new report, technical experts from the Brookhaven National Laboratory would actually develop the study, as they had in the WASH-740 report.

If any of the men beginning the new study that spring could have foreseen the problems they were going to face, or what was going to happen to their painstaking efforts, they might never have begun it.

NINE

William Lyons Phelps once said that great literature is never written by committees. The rewrite of WASH-740 was not destined to be an exception to the rule.

Cliff Beck headed up the steering committee of nine or ten scientists from various AEC divisions that represented a confusing array of departmental acronyms. Ken Downes of the Brookhaven group was project director for the academic task force that would be doing most of the spadework on contract for the AEC. The Brookhaven group would be working mostly in their Long Island laboratories, away from the day-to-day traffic of the AEC Washington offices. They would be a semi-independent group under the somewhat jittery eyes of their official AEC colleagues.

Cliff Beck launched an early meeting in August, 1964, with a rundown of what he hoped to accomplish. The new report would have to skate on a thin line, he told the committee. It would have to avoid the twin pitfalls of overpessimism, which could severely bruise public acceptance of nuclear energy, or underpessimism, which would look like a whitewash of the atomic power plant dangers. Most of the committee agreed that, because the possibility of a catastrophic accident could not be ruled out, a major job was to estimate the maximum damage to people and property that would be created by a runaway meltdown of the newer atomic power plants planned to spread across the country in the early 1970's.

There were two clear areas that would have to be carefully studied. One was the estimates of death and destruction from a major accident. The other was the *chances* of such a major disaster happening. The August, 1964, meeting did not neglect this latter point which posed some difficult problems.

Ken Downes and his scientific group from Brookhaven were convinced that nuclear power stations hadn't been around long enough for them to make any rational kind of probability estimate. This point created an immediate hairline crack between the philosophies of the AEC and the Brookhaven groups. AEC's Dr. R. L. Doan said bluntly that there must be a compromise between the scientist's desire for nice calculations and an approximate grasp of the real probabilities. Ken Downes, however, felt that the assigned job was to discover exactly how much damage would result from a catastrophe, regardless of the chances of it happening.

There was a muffled edginess developing between the two groups. Already, there was a blizzard of interoffice memos stacked in the in-and-out baskets of both Brookhaven and Washington desks. It was determined, however, that the report would cover basically the new, large light-water reactors, considered safer than the breeder type being built on Lagoona Beach.

F. P. Cowan, a Brookhaven health physicist, passed along the gossip on the early August meeting to his boss:

> The general guidelines for the project were decided on. However, they don't plan to meet again until the end of September (everyone is goofing off to Geneva) and, since the final rough draft

is due October 31, the Steering Committee won't do much steering. . . .

They shuddered at the idea of treating cases of a reactor located in a city, but didn't forbid it. Since conditions anticipated during the next decade are to be considered, it is almost certain that we must consider cases where reactors go to populations or vice versa. . . .

What Cowan didn't mention was that *all* atomic power plants would have to be located in or near a city, in order to make them economically possible. The high cost of transmission lines would make them prohibitive in cost otherwise. Yet, to even consider an accident at an atomic power plant in a heavily populated location was enough to make the most sanguine scientist shudder.

Radiation inventory increased with each generation of reactors. Cowan became more concerned with establishing what radiation doses would be lethal, and figuring what changes might be made in the old study. He noted: "First paragraph is pretty pessimistic, and we might make it a little more favorable this time. . . . If we wished to refine matters a little, we could use 100 [rads] to 300 as illness, 300 to 600 as half deaths and half serious illnesses, and over 600 as fatal." Scientists varied in their views of the exact lethal doses, but 450 to 500 rads were generally considered enough to kill half the people exposed to it.

His other notes were less formal: "Find out from Tompkins if F.R.C. is planning any Strontium-90 mischief. . . . In general, our Strontium-90 reasoning is too casual to pass muster at this time." And, in regard to a radiation "dose to the gut," he wrote: "Unfortunately we probably can't get away with the crudities of WASH-740 even though the nature of the project probably justifies them. . . ." He was referring to the damage Strontium-90 could do as a residue in milk or crops, and similar damage by radiation absorbed in the digestive system.

To the layman, this casual bandying about of such commentary might seem somewhat chilling. It could be especially so in referring to the potential radiation damage to the very young. He mentioned that one of his colleagues was ". . . influenced by the widely publicized claim that thyroid cancer in children has occurred at levels of exposure of 150 to 200 rad. In considering infants, the above value of 170 rad would be satisfactory, but the intake to produce this dose will be one-tenth that for adults. . . ."

To nuclear scientists, however, this was a job to be faced and done. It was the climate in which they lived and worked. They could not stop and think too much, like the layman, of the implications of such a disaster, or it would render them ineffective in their routine, just as a virologist faces deadly viruses every working day of the year without fear.

By its very nature, Cowan's job as a health physicist could never bring much comfort to the average citizen. Radiation of any kind was simply not a pleasant thing, and never could be, regardless of the one AEC attempt to refer to it as "sunshine units." Its end result, always and without fail, was the ionizing of the atoms in the body, a process the body atoms were never built for. If a molecule in the body is hit directly, it becomes a biological cripple, and worthless. Other irradiated molecules can enter the body indirectly and can seek out the DNA—the blueprint or computer that tells the cell what to do—with the same results.

The stuff Cowan dealt with for the study was of this nature. He broke down the hazards into two groups. One was the indirect hazard that ionizing radiation had on the trees, crops, milk, land, homes, and the general environment. The other was the direct hazard—the irradiating of humans.

Cowan noted in his report for the AEC committee that the problem of a fission product released from an atomic power plant was much simpler than that of a nuclear bomb. The release from the power plant probably would not be complicated by a major blast or fire. But still there were complexities.

Certain vegetation could die in a matter of days from relatively low gamma exposure. Other fallout from an atomic power plant of some 10,000 rads was enough to kill the dominant trees of the deciduous forests. Lower exposure could kill most pines and evergreens. "This means," Cowan said in his appraisal, "that gross changes in natural ecological systems can be anticipated following a reactor release whenever the 3–4 days exposure to the general environment is in the range of thousands of roentgens. . . . It seems probable, however, that most crop plants exposed to such high levels of radioactivity from a reactor release would be contaminated to the extent that they could not be decontaminated and they would constitute a total loss."

Cowan went on to consider that the principal long-term hazard to man would arise from the internal "emitters" absorbed through the food chain—the damage from Iodine-131, the bone-seeking Strontium-90, and Cesium-137. About the only possible "treatment" would be to abandon the land or rip up the vegetation and deep-plow the soil that was left. But he was concerned about the lack of adequate data on all these poisons.

This was to be a constant problem throughout the study: the lack of sufficient information. The damage to a one-year-old child from inhaling radioactive iodine, for instance, could only be guessed as being three times as bad as that for an adult. Allen Brodsky, a research associate from the University of Pittsburgh, wrote Cowan that he was having trouble with what was, or what was not, a safe limit for various doses of the fission products from an atomic plant accident. He ended a letter to Cowan: "Wish me luck in defining 'acceptable.' "

From a different vantage point, Stan Szawlewicz of the AEC's reactor development division notified his boss that the new study was being put together to help solve the insurance question, and also to overcome the obsolescence of the tired

WASH-740. He mentioned in his memo "the fond hope that some of the pessimism reflected in the WASH-740 models could be reduced by new information." He hoped now that improved calculational techniques, and discretion in selecting the type of accidents, would reduce the awesome figures of the old 1957 study.

Szawlewicz seemed a little edgy about Ken Downes. He found Downes unwilling to change or admit the need to change the basic accident assumption in the new big reactors on the horizon: that they carried with them so many more poisonous fission products than the smaller reactors in the old study. He was worried about Downes's choice of the worst conceivable accident to study, where all the fission products would be released.

He had other concerns, similar to Cowan's. Reactors in or very near cities would be a problem. The study of other than the light-water reactors was left open. Reactors such as the Fermi liquid-metal fast breeder were a problem ". . . since the report could very easily show unfavorable safety comparison between reactor types to the detriment of the industry. . . ." He was disgruntled at Ken Downes's selection of such a big sample accident, since it could create uncertainty and disagreement, and would dominate the meetings.

There were other snowflakes in the blizzard of memos. Jim McLaughlin, an AEC radiation physics man, wrote to Allan Lough of the biology and medicine division: "The A.E.C. should not place itself in the position of making the location of reactors near urban areas nearly indefensible. . . ." He was interested, too, in what the odds were of an accident happening.

McLaughlin's rough figuring showed that the chances based on experience to date would be three accidents in ten years, when the 1,000 planned reactors were built across the country. Others went along with this estimate, although everybody agreed that no one could reliably make any probability estimate now or in the immediate future. There simply wasn't enough data to go on.

Dr. Bernard Pasternack, a consulting NYU biostatistician, basically agreed with McLaughlin's estimate, stating that three or *more* accidents among 1,000 reactors in ten years was virtually a hundred percent certainty. Pasternack also warned about progressive changes in a reactor's condition. Over the months and years, parts would get swollen and worn out from the intense radiation inside their guts. There was little data to go on in this respect. The Brookhaven team continued to insist that it simply would not work on the probability picture. They threw the problem back to Cliff Beck, with the implication that it would be charlatanism for anyone to make a prediction on the possibility of a disastrous accident in a reactor, without a long history of reactor operation to base a prediction on.

The memos continued to pile up, none of them reflecting that much coherent sense could come out of the study. M. E. Smith, a Brookhaven meteorologist, found his probe into available facts on weather conditions in the case of fallout extremely unfruitful. He found that many statistics were "pretty uncertain," "almost unbelievable," "largely useless," "described in uncomplimentary terms," and concluded: "I don't quite see how anyone really knows."

The poisonous fission products released within and on the boundaries of cities was a ticklish subject, because of the high potential of deaths, and destruction to a large population. The tendency of some of the committee was to push this kind of problem under the rug. But Smith disagreed. He told his colleagues this danger should be studied, "owing not only to the fact that this is likely to occur within ten years, but also that many people consider city reactors desirable as an alternative to sulfur dioxide pollution."

Brookhaven's I. A. Singer felt that nearly everyone basically agreed, without liking it, that the reactor near-or-in the city had to be considered. Singer noted: "Don't really know what the leak rate will be in any given accident, since the shielding deteriorates in time." If there was deterioration at any point in the reactor vessel, pipes, or containment shell, the damage, of course, increased.

The heat from the meetings in the summer of 1964 had barely cooled when another major meeting was held on the fresh, green, campus-like grounds of the Brookhaven Laboratories in October. By now there were definite straws in the wind that the AEC steering committee in Washington was getting somewhat liverish about the relative academic purity of the Brookhaven working committee.

As the head of the Brookhaven committee, Ken Downes opened the October meeting by saying that his group had been asked to answer the question about the maximum damage that could result from an atomic power plant accident. In light of the many AEC memos expressing opposition to Downes's sample accident case, this was like putting up a red flag at the very start. It gave an immediate signal that there wasn't much hope for a better picture than the one that the WASH-740 report had projected.

U. M. Staebler of the AEC reactor development division jumped back to the question of why the Brookhaven people refused to work on the probability question, which at least might show that the chances of a catastrophic accident were very slim. Again the answer was that any statistician who proposed such a scheme in the light of such meager data would have to be a "fringe member of the statistical community." An attempt was made by some of the AEC group to see if it could be proved that a hundred percent melting of the core was impossible. Brookhaven contended that there was no basis for this. The after-heat was enough to complete the entire melting of the core and to produce the resultant catastrophe if the containment was breached. To analyze the "maximum damage" sample case, Brookhaven had arbitrarily assumed an opening in the containment building the size of a doorway.

This, their computations showed, was enough for all the fission products to escape. It would take about an hour for half the core to melt; a day for all of it to melt. If a city was involved the results would be catastrophic, and there would be deaths as far as seventy-five miles away from the atomic power plant, with destruction far beyond that. Not even the meteorology figures could change the end results to any measurable extent. If the wind moved out, it would reach further out to kill. If it hung still, it would kill more people in the city. "A catastrophe still results," Downes told the meeting.

It was a pessimistic picture, and it got worse as Downes went on. The results his computer had come up with in the new study were "very horrible," he confirmed—far worse than the WASH-740 estimates:

> instead of 3,400 deaths, there would be 27,000.
> instead of 43,000 injured, there would be 73,000.
> instead of $7 billion in property damage, there would be $17 billion.

In other words, the death figures for the new reactors—with their larger fuel loads and higher power—had risen to eight times the original toll. The injury figures had almost doubled. The property damage had more than doubled. The burning question was, and would continue to be: What would happen to the entire atomic power plant industry if these figures were released to the public?

Downes admitted that the results were "frightening." Dr. Winsche, Downes's superior, verified this by adding that unless some "mechanism" could be found to make their assumptions impossible, "the numbers looked pretty bad." As one final lunge at optimism, someone asked Ken Downes if he had taken the evacuation of a population from their homes into account.

Downes said that this had been considered in their computations, but was found to make little difference in the results.

Dr. W. D. Claus of the AEC division of biology and medicine made the understatement of the day when he said that he was concerned about the direction the meeting was taking. It appeared to him to be a mixture of arbitrary—although possibly expert—choice of accident characteristics, with practically no basis for attaching a money value to a "holocaust." (The word "holocaust" was eliminated from the official minutes of the meeting.) He would like to see various types of accidents related to their probabilities so that liability experts could put dollar estimates on them. Dr. Kruper reminded him that this had already been done in the rough figures by Jim McLaughlin, and they were terrifying: three accidents in ten years when 1,000 reactors were completed. Even though these figures were provisional and unscientific, they held out no promise for more cheerful results when more operating experience was available.

The major difficulty seemed to be that no one could rule out a nuclear catastrophe, although the AEC group was doggedly convinced that it would not happen. They had nothing to base this confidence on except that a catastrophe hadn't happened yet.

On this note, the meeting adjourned for a lunch break, but the AEC group under Cliff Beck lost no time in gathering for a private huddle of their own. The AEC members tried to redefine just what the purpose of the report should be. To get around a full scientific exploration of the tricky probability question, Beck suggested that a whole section could be devoted to that. The AEC group, however, all agreed that the probability question should be developed and answered, in spite of Brookhaven's obvious slurs about shoddy statisticians who would accept such an assignment.

David Okrent, the one AEC man to have come in from Chicago for the meeting, told the group that it appeared to him that the committee was unhappy about "the catastrophic results of the figures." He said it seemed that they "secretly hoped that some other group would supply optimistic probabilities which could be applied." This evasion of facts disturbed him very much.

Okrent, who had written a textbook chapter on breeder accidents with Walter McCarthy, came to the defense of what Ken Downes had said by pointing out that, when anyone looked back at the former hazard reports, it was obvious that those accidents that were once considered incredible could now be considered credible. Therefore, Downes was right if the objective of the study was to show the worst that could happen.

What worried AEC's Doan most was that the catastrophic figures in the new study would "strengthen the opposition to further nuclear power." But Beck countered that the results of the new report "could not be ignored just because they were unpleasant." Doan agreed with Beck that they would have to put more into the study of the probability odds, and how this would affect the engineering safeguards. Without this, they felt, the report would not be appropriate for release to the public.

By the time their off-the-cuff meeting was over, they barely had time to grab something to eat and rejoin the Brookhaven group for the afternoon session. Cliff Beck explained the conclusions of the closed AEC meeting: That the first section of the new report would have to deal with the probability question, in spite of Brookhaven's objection that such a study would be meaningless. The rest of the new report would deal with the consequences of accidents in terms of the different types of fission products. It would also cover the type of accidents, and would include the minor ones as well as Brookhaven's estimates for the big, catastrophic accident. They would shoot for a first draft to be completed by December.

By now the scuttlebutt about the pending report had provoked a tremendous amount of curiosity, interest, and apprehension—especially on the part of the public utilities, reactor manufacturers and suppliers. A bad report could be ruinous to the nuclear power industry.

The industry group began flooding Cliff Beck with offers to help in the study. Beck had been parrying with them. But he did feel that the first draft of the report should be given to the safety committee of the Atomic Industrial Forum for review. This organization consisted of every important corporation involved in the atomic energy power plant industry. In addition to the obvious ones, like General Electric, Westinghouse, Babcock & Wilcox, and other reactor builders, it included many universities and foreign companies. Walker Cisler had been a founding father, first president, and a director of the organization back in 1952. Its more than six hundred members made up the guts of the nuclear industry.

There was no doubt that the reaction of the Atomic Industrial Forum to the figures that were now being developed by the Brookhaven experts would bring enormous pressure on the two committees studying the problem, but everyone agreed that the Forum should be consulted. Dr. Doan felt that it would be interesting to find out what an outside group like the Forum would suggest for the new report. He was convinced that it would be very difficult for them to come up with concrete proposals. Like their own group, they would have to resolve the dilemma that hung over their heads: They needed the Price-Anderson Act insurance—but at the same time it was "difficult to prove that reactors are safe enough to build," as Doan put it.

The meeting came to an end on this note, with Cliff Beck's final wry comment that "the report may surprise some people." There was no question that it was likely to. Some facts that emerged in the discussion were particularly disturbing. For example, the new reactors would contain 50 to 100 times the amount of Strontium-90 in an atomic bomb; in fact, the bomb was somewhat less dangerous to a local area as far as this was concerned, because the bomb blew strontium straight up in the air. The reactors would spread it silently along the ground.

For Stan Szawlewicz, the study had taken a critically bad turn. As chief of AEC's research and reactor development, Szawlewicz was motivated by the strong desire to push the reactor construction program forward at all possible speed. He revealed that he was distressed by the very pessimistic results of the calculations. He was also disturbed by the dangers of publishing the new report as the facts were emerging.

Szawlewicz was convinced that those who opposed the construction of new atomic power plants in their cities or towns would seize on the disaster figures and ignore any qualifying statements. He felt that experimental proof of the effectiveness of the containment vessel was difficult and expensive to achieve. He wrote in a memo reviewing the October meeting: "The results of the hypothetical B.N.L. [Brookhaven National Laboratory] accident are more severe than those equivalent to a good-sized weapon, and these correlations can readily be made by experts if the B.N.L. results are published." He also noted that the Brookhaven figures could easily be applied to the moderate-sized reactors being built, which would show the consequences to be almost as bad as the bigger plants. "This might have serious consequences in obtaining site approval for such reactors," he wrote.

Szawlewicz suggested that perhaps the ground rules for the study could be changed, so that only those accidents that did not breach the containment vessel would be presented. He felt that the steering committee should meet and discuss what the new report would do to block the whole progress of reactor development and construction, *before* publication, not after—when it might be too late.

He was joined in this conviction by Howard Hembree, also of the reactor development division of the AEC, who suggested that corporations like the Phillips Petroleum Company be brought in to help steer the Brookhaven scientists back onto what he and Szawlewicz thought was the right track.

Hembree called Ken Downes from Washington as a follow-up to the October meeting, and began exerting pressure to have him bring the nuclear manufacturers and private contractors into the arena. Downes obviously didn't like the pressure from Hembree, who was not even a member of the AEC steering committee. When Hembree offered him the "assistance" of the Phillips Petroleum Company, Downes replied stiffly that the company was available to him through his own direct contacts, and it was also within his authority to ask industry to review his work at any time. But Brookhaven did not want to do this until it had clarified its own thinking.

Further pressure wasn't long in coming. Within a day of Hembree's phone call to Downes, Bill Cottrell and George Parker of the Union Carbide Company, which operated the AEC's Oak Ridge National Laboratory on contract, descended on Ken Downes to express their disappointment with the ground rules of the WASH-740 update. The Union Carbide men were anxious that Downes should include some of their own research on fission product behavior. In a letter following his visit, Cottrell wrote to Ken Downes:

> Although we have no responsibility for your study, we are vitally concerned with the results and the impact that it will have on the nuclear community. Our disappointment in the ground rules for the study is mild relative to our apprehension regarding the publication of the results of a study with such narrow objectives. We know from experience as well as association that difficult concepts such as those to which your study is addressed are readily misinterpreted. Furthermore, there are many who would knowingly distort the information to further their own interests. While the small technical group for which the report is intended would certainly apply the results properly, the study's very existence originated from a public interest so that sooner or later, the study will be subject to public scrutiny. . . .

> We feel quite strongly that your present study will be subject to much misunderstanding and misinterpretation and will have a net result that will be quite detrimental to the exploitation of the potential benefits of nuclear science and technology. Accordingly, we would not wish to be associated with the report and request that you do not acknowledge our assistance in any manner. . . .

It was only one of many memos, letters, and discussions to follow, nearly all of them expressing an overriding fear of what would happen if the results of the study were made available to the average citizen or critical scientist.

Szawlewicz continued to be the AEC bird dog of the opposition to the Brookhaven stance on the report. He wrote to Staebler on November 27, 1964, to say: "The impact of publishing the revised WASH-740 report upon the reactor industry should be weighed before publication. A very strong section on the low probability of accident occurrence must be presented to counteract the expected effects of the hypothetical accident calculations."

Szawlewicz also felt that if the catastrophic accident showed even small odds on happening, entirely new post-accident counter-measures should be developed. These would include remotely operated machines or robots for closing doors or valves, or bulldozing earth-fill against the breach of containment.

At a special meeting of the Phillips Petroleum men with other AEC reactor development experts, pessimism still ran deep. An Oak Rdge representative felt that the only thing to be done was to urge that "the report not be issued in its present form." The upshot of the meeting was summarized as:

1. Persuading the AEC steering committee to steer the issuance of the report in a "reasonable direction," and
2. A double check by Phillips of the Brookhaven figures in the hope that some factors had been overlooked that might reduce the disastrous figures that had emerged.

Meanwhile, Cowan, Brookhaven's health physics expert, was groping to find meaningful estimates of the amount of radiation exposure the public could take. It was important to establish the maximum dose of radiation that would trigger the necessity for immediate evacuation of people from their homes. "Some areas," Cowan wrote, "will be contaminated to levels such that external exposures due to fallout will be very high and, hence, a figure for urgent evacuation is needed. Here, the action is necessary to forestall exposures that are potentially genuinely harmful. Thus, I believe that 25 rads in one day would be a reasonable figure." (Since a "safe limit" for an atomic worker was 3/10ths of a rad for a week's duration, Cowan was being very generous in his allowance.) Cowan also noted that the eventual figures set in England after the Windscale accident was 20 rads for children and pregnant women, 30 rads for others, and up to 60 rads for a limited group of workers.

Fear continued to grow about the results of the study getting out to the general public. Corporations who were working on contract for the AEC were especially concerned. A letter from Union Carbide to Hembree typified this fear: "Perhaps it may be worth noting that the initial reaction of ORNL [Oak Ridge National Laboratory] was one of disappointment in the choice of ground rules, and also one of great apprehension regarding the publication of the results of a study with such narrow objectives." It was hard to define what was meant by "narrow objectives." The main concern of industry was simply that the figures were so overwhelmingly staggering that a storm of protest was sure to arise if they got out to the public.

While the internal struggles were going on in Washington and Brookhaven, troubles continued to boil all through 1964 at the Fermi site, although the reactor was merely going through tests. An oscillator rod disassembled, the number 5 safety rod tube was damaged, sodium pumps failed, the enormous cask car "locomotive" was continually breaking down, the steam generators often leaked and failed to work properly, welds cracked, the number 1 safety rod dropped unexpectedly and its bellows leaked, and many leaks were found in the operating floor seals of the containment building.

With the fast breeder reactor being promoted as the hope for the future, such troubles did not augur too well. As far as the AEC-Brookhaven study was concerned, the breeder—potentially the most dangerous of all reactors—continued to be ignored in its estimates. There was enough trouble with the light-water reactors, which had a far more simple design, and were somewhat more predictable.

The entire new study was a hypersensitive matter. One AEC commissioner, John G. Palfrey, addressing the Atomic Industrial Forum in California at the beginning of December, 1964, was supposed to discuss it with the industry group. Instead, he said: "It would be possible for me to begin and end my speech in 30 seconds. . . . The Brookhaven study is not ready yet, and the Price-Anderson [insurance] report is still in draft form, and has not been reviewed by the Commission. . . .

"I think I should beg off the Brookhaven study because my guesses on what it will say or should say could be irresponsible chatter on a ticklish subject that needs accurate reporting."

What Commissioner Palfrey didn't realize was that by bringing the new study up so glaringly in the public eye, he was digging himself and the AEC a hole that later on would be very difficult to get out of.

Meanwhile, more meetings of the Brookhaven and AEC study groups did nothing to cheer up the situation. In fact, much of the discussion centered on how the gory details of an atomic power plant accident could possibly be released to either Congress or the public.

Advertising by the utilities had painted a lovely picture of pollution-free power plants that were paragons of virtue. Promises of absolute safety to the public had been lavishly presented in newspaper, television, and radio advertisements. A bright new era of power was being presented with all the consummate skills of Madison Avenue. The average citizen, without detailed knowledge of what went on inside a nuclear reactor, was being lulled into a feeling of euphoria, while inside the AEC and Brookhaven conference rooms there was casual discussion of contamination that could cover an area as large as the state of Pennsylvania, kill 27,000 people at one crack, and knock out a perilously high percentage of the Gross National Product.

It had all begun so quietly. The congressional Joint Committee on Atomic Energy had merely asked the AEC for a report so that the committee could decide whether it could get the government off the hook on the insurance problem. Now the figures were coming out so horrendously that the AEC was hoist with its own petard. The plans for the study had been announced publicly. There was no way to back down.

The meetings continued, with endless haggling. The scientists reviewed how an accident would begin: with the melting at the center of the fuel rods, then with the stainless steel grid and cladding drooping to a messy puddle that would drop to the bottom of the reactor vessel while the operating crews stood by, helpless to stop the disaster. Then the hot, uncontrollable fuel would spall—chew up—the concrete base. Most of the damage would come from the radiation released in the first two hours, leaving two or three hours for evacuation. In homes near the accident, Brookhaven chemist Anita Court noted, there was very little that could be done; the doses were higher and would be received sooner. With a large city population, evacuation would be difficult, if not impossible. Ken Downes commented that even shelter would not be helpful. The air turnover rate in a house was high. There was no natural mechanism to reduce the amount of fallout released. Nothing that had developed in reactor design or construction since the 1957 report could alter the results.

Cliff Beck made it clear that he would like to avoid publishing the actual figures that Brookhaven had come up with. But Cowan immediately asked how this could be done without the Joint Committee on Atomic Energy, or anybody else, knowing that many of the results would be 50 to 100 times worse than in the old report.

David Okrent summoned the courage to say that there was no alternative to a public document. He felt that the report should be done "without glossing over the fact that the consequences are worse." This in turn might lead to considering the possibility of underground locations for reactors.

This concept reflected Edward Teller's conviction that no atomic power plant should be built above the ground. But the costs would be substantially higher. Okrent was intent on the premise that the AEC commissioners should get the data so that they would realize what they were facing in the way of such potential damage.

The one hope that seemed to predominate at the meetings was that if the probability, the odds, the chances—whatever one called it—could be shown to be small, the impact of the numbers on the public mind would not be so great. Cliff Beck suggested that even if they couldn't specify definite numbers, they could say that "the probability is believed to be extremely small."

The ultimate upshot was a decision to have Brookhaven work up a draft of the report, specifically marked "For Official Use Only." But the main and knotty question still dangled in the wind. As Szawlewicz put it in his own notes: "The results of the study must be revealed to the Commission and the J.C.A.E. [Joint Committee on Atomic Energy] without subterfuge, although the method of presentation to the public has not been resolved at this time."

Hardly any member of either the AEC or Brookhaven group failed to reflect a sense of futility. Like the Fermi scientists, they too were men of goodwill, sound intellect, and with a sense of moral obligation to themselves and to society. But they had opened Pandora's box, and were now faced with the inevitable problem of what they were going to do.

A small hope suddenly emerged when it was discovered that a company known as the Planning Research Corporation was already undertaking some probability studies for a different AEC division, and would be able to temporarily switch over to the reactor problem. The company, a California-based research group, was not considered a "fringe member of the statistical community"; they freely admitted that there was the lack of data to work with, and agreed to settle for "quasi-quantitative" predictions, rather than precise ones. With considerable hope and interest, the committees awaited the first attempt by Planning Research Corporation to put a definite figure on the odds of a reactor accident. It just might determine whether they could issue the new report to the public; or whether another atomic power plant would ever be built.

TEN

At the beginning of 1965, the Planning Research Corporation sent its first working paper to Cliff Beck. About the only thing that the research group could report with any degree of accuracy was based on the 1,500-reactor years of experience that had been achieved up to 1965. A "reactor year" figure is simply the number of individual years that all the reactors in the country had been running. It would be the same as 150 reactors running for 10 years each. Or 300 reactors running for 5 years. Or 500 reactors running for 3 years. By about 1985, there would be 500 reactors, the projected figures showed.

Using a complicated method that assumed that catastrophic accidents would happen according to random tables, the results of the study turned out to be something just short of horrendous. The report showed:

> We are 95 per cent confident . . . that the probability of occurrence of a catastrophic accident during a reactor year is less than 1 in 500 . . .

This figure would mean that in 1985, when 500 reactors

were spread across the country with at least one in almost every state, there would be the possibility of one major catastrophic accident *every year.* And when the AEC reached its goal of 1,000 atomic power plants, the possibility would rise to one major holocaust somewhere in the United States *every six months.*

Furthermore, the figures considered only a meltdown accident. They did not take into account an accident that could be triggered by an earthquake, or an aircraft or a missile or even a meteor hitting the reactor facility. Earthquakes were a very real factor in California. Nor did the researchers consider transportation accidents with irradiated fuel assemblies. In transporting these for reprocessing, the assemblies were packed in containers that could only withstand a fall of 30 feet, a collision of equal impact, or a 30-minute fire at 1400°F. The question was: What about a drop of 35 feet, or a 40-minute fire? Although control room operator error was considered, nothing was said about psychotic behavior in the control room, or the seizure of an atomic power station, or stolen plutonium that could be made into a bomb.

But whatever the situation, the figures were totally unacceptable. The Planning Research Corporation was aware of this, and had already developed a method for taking the curse off their computations. First, they emphasized that the chances were *less than* 1 in 500 reactor years. Then, because of the paucity of data to go on, they relied on a systems engineering analysis that was, as they called it, "probabilistic in spirit" and based on the "judgment of experts." In this way they could set up what they called certain "chains of events" that had to occur before a catastrophe took place. Through this device they were able to suggest that the odds against a catastrophe were something like 1 in 100 million, or even less.

No one, not even the researchers themselves, felt that either the low or the high estimate were realistic. They admitted the dilemma in the report for Cliff Beck, in which they pointed out that "We simply do not have enough data."

Brookhaven was having its own difficulties in trying to draft a low-key statement on the results of a catastrophic accident. Phrasing was important, and the search for euphemisms was obvious. The Brookhaven writers tried to be frank, but always ended up with clumsy attempts at smooth and tranquil words. Their memos and reports constantly reflected the conflict between what they felt was scientifically accurate, and what the AEC and industry pressure was forcing them to consider. On some points they would give in; at others they would balk.

The introduction of the new Brookhaven report pointed out that every nuclear reactor creates radioactive nuclei as a by-product. It went on to say that fission products, "if released to the atmosphere in large quantities could cause damage to the general public. A major effort of the nuclear industry and the Atomic Energy Commission is pointed toward preventing such a release. Thus far, efforts have been successful but it is recognized that there is no absolute guarantee that this will always be the case."

The basic conclusion of the old, 1957 WASH-740 report was reiterated:

> It could not be proven that fission products would not be released as a result of a major accident in a nuclear power plant, although the probability of a large-scale release was believed to be extremely low.

The conclusion of the new study was almost piquant: "The result of all this work can be summed up with almost distressing simplicity." They saw no reason to believe that the extent of damages would ever "be less than those estimated in WASH-740."

This statement would not make supporters of nuclear power plants stand up and cheer, especially when it was followed in the body of the report by a sentence saying that, in examining the most pessimistic accident, all the facts lead "to enormous potential damages. . . ."

The report continued in a suffocating but honest cloud of gloom. As far as estimates of the liability for damage was concerned, they would come to "an appreciable fraction of the Gross National Product of the United States." The report went on to describe the way the fuel would melt in an accident. This was followed by the description of how the fission products could escape in a "Gaussian plume"—the pattern revealed in visible smoke as it fans out into the atmosphere from a chimney. In a fission product release, of course, no one can see the plume; it would simply be there.

The report showed how weather could affect the release. Under certain conditions during the day, less than 200 people might receive a 1,200-rad dose, more than twice the lethal exposure. But the same release at night from a reactor in the center of a city might fatally injure about 45,000 people.

The figures showed how the personal and property damage could cover as much as 600,000 square miles, with contamination by Iodine-131, Strontium-90, Cesium-137, the rare earths, and other fission products. The fission products were broken down as to what they would do to the whole body, the bone, the thyroid, the lungs, the skin, and the gastrointestinal tract. Monetary estimates of damages played an important part in the draft too, stressing the likelihood of $17 billion worth of destruction. This aspect, in fact, was too heavily emphasized for even some of the Brookhaven men.

M. E. Smith and his associate I. A. Singer, both of the Brookhaven group, wrote to Ken Downes that there was some important alterations they wanted to see made. What bothered them most was the emphasis on material damage: "Both of us reacted most unfavorably to the cold, hard-cash tone of the draft. All of us understand what the Commission had in mind in having us review this problem, but we do not see why Brookhaven's report needs to be couched solely in monetary terms. . . . What we are really trying to say is that we still find that a large release could result in a major catastrophe and an enormous area contaminated. I suggest that we remove the dollar comments entirely, except perhaps in the historical section. . . ."

On the AEC side, the working draft of the probability section was undertaken by Cliff Beck personally. It, too, opened with a reference to the old WASH-740 report, whose ghost seemed never to die. Beck reiterated this statement regarding the "exceedingly low" probability of a catastrophic reactor accident: "One fact must be stated at the outset: no one knows now or will ever know the exact magnitude of this low probability of a publicly hazardous reactor accident." He followed this up with: "It still must be admitted that the theoretical possibilities for such major accidents do in fact exist." Then he went on to quote the Planning Research Corporation figures, and to say that past history had shown a favorable indication of the stability of reactors.

Beck's draft added, however, that "there have been discovered incipient failures which would have to be classified in the category of 'near-misses' of serious accidents. For example, in three reactors, two or three of the stud bolts on the head closure of the main pressure vessel were found to be cracked from stress corrosion. Failure of a sufficient number of these bolts would have resulted in the head being blown off, with high probability that the force would penetrate the walls of the containment building.

"In another reactor two main control rod shafts were found to be cracked from stress corrosion," Beck's report continued. "In three reactors significant cracks were found in the piping of the main primary coolant system. In another reactor, four of the stud bolts which held the bonnet in place on the main primary system flow-control valves were found to be cracked."

He concluded his draft with a reference to the possibility of a big, disastrous accident: "The possibility that such accidents might occur cannot be excluded, and there has been accumulated some evidence that a few failures may have almost occurred which could have resulted in more serious accidents than any which have thus far been experienced. . . ."

On January 19, 1965, a new meeting of the AEC-Brookhaven groups was called. It was only a warm-up for a meeting scheduled with the industry group of the Atomic Industrial Forum to be held later toward the end of the month, but the intramural meeting was heady enough. It began quietly. Cliff Beck praised the excellence of the Brookhaven draft report, indicating that it was close to the target. He was sure that the industry group would be convinced that there was no real way to reduce the damage figures, and that they would concentrate on the probability section to give "the best possible perspective."

The meeting had not moved along very far before the nitpicking began, in spite of Beck's high praise of the Brookhaven draft. Dr. Doan objected to the fact that engineered safeguards had not been factored into the report, and felt that people would say: "Look how far we have come with engineered safeguards, yet look how big an accident we get." He also went right to the point that the results of the Brookhaven part of the study could be summed up "with distressing simplicity." It was the only colorful comment in the report, and this phrase said a lot in a few words.

There was a great deal of fishing in trying to find ways to soften the impact of the casualty and destruction figures. For example, might a nuclear bomb be more likely to go off and wipe out a city, than an atomic power plant? Dr. Doan suggested that they pick out some huge plant making a noxious product, and compare the damage from a major accident in such a place. Szawlewicz came up with Fort Detrick. Ken Downes joined in by offering the comparison with a nitrate ship that could blow up in a harbor. But Cliff Beck thought such ideas would be questionable.

After a hasty lunch, the discussion got back to the probability figures from the Planning Research Corporation. Dr. Doan said he felt that the reader had a right to expect something more than futility from the report, that it should indicate what had been done to prevent accidents, and thus lift the "pall of gloom" from the situation. Then he added that if nothing reassuring could be said, perhaps they should stop building reactors.

It was generally admitted that there were still major problems in the construction of the atomic power plants. Brookhaven's Dr. Winsche emphasized that there was great confidence that a given accident will not happen, but they couldn't assure the public that it would not.

The official minutes of the meeting continued with: "Dr. Beck noted the problems of radiation embrittlement in steel and the number of defects already found in just one reactor, with a great number of reactors just like it. Thus, it is not at all assured that the conditions assumed in this report cannot happen. He felt that we cannot predict if, or when it might happen."

Beck punctuated this by saying that the figures brought together by the consulting research team had to be taken with a grain of salt. He felt that all that could be said was that the probability is extremely low. One of the Brookhaven men said that this was the reason why Brookhaven hadn't wanted to tackle the job. The minutes revealed the subtle difference in attitude between the AEC and the Brookhaven men.

The afternoon session turned to the subject of "near misses." Szawlewicz argued that if these were covered, "care should be taken to avoid implying that a catastrophe could have followed."

One of the things that concerned Cliff Beck was the statement that kept cropping up in the various drafts. He felt that they could not keep repeating on every other page that "the probability is very low," or the accident is "highly improbable."

Since the report frankly recognized that serious accidents *are* possible, it should not claim that they cannot happen, even though there was confidence that they wouldn't. Doan agreed. He was concerned that there was a possibility of some of the critics picking this up and taking legal action against further construction of atomic power plants, on the grounds that the new report showed that the AEC was being irresponsible in granting licenses. Along this line, someone suggested dropping the term "near miss," and Beck went along with the suggestion.

The meeting dwindled to an end by 3:25 that afternoon. There was at least one thing certain: The first draft of the AEC-Brookhaven study would be stamped "For Official Use Only."

Szawlewicz might have been contentious, but he knew how to put his finger on a problem. In a report to his superiors he boiled it down to this: (a) If the engineered safeguards are assumed to work, then there is no public liability problem. (b) If they are assumed to fail, then it is difficult to describe an accident level that represents a true upper limit for liability purposes.

But this was partially answered by Cliff Beck's "Official Use Only" draft of January 21, 1965. In it, he said: "In any machinery as complex as a reactor facility, it is inevitable that structural failure, instrument malfunctions, operator's efforts and other mishaps will occur, despite the most careful design and rigid schedules of maintenance and administrative control. Such has been the experience with reactor installations."

According to this, engineered safeguards could not always be assumed to work. And therein lay the rub. Beck's new draft emphasized this by adding: ". . . despite all precautions, a complete assurance against any such possibilities can never be fully established." The draft also pointed out that ". . . simple mathematics would show that the probability of catastrophic accidents, at the 95 per cent level of confidence, is not more than 1 in 500 reactor years of operation. The actual probability is certainly a great deal smaller, but the extent of reactor operation is simply insufficient to demonstrate this."

The papers prepared by the Brookhaven and AEC groups were dutifully sent out in draft form to the members of the industry group that made up the representatives of the Atomic Industrial Forum. There was statement after statement in the drafts that would not exactly charm the industry men. Such phrases as "the emergency core cooling system cannot be made foolproof," or, "conceivably the damages could be substantially greater" than in WASH-740; or, "it cannot be assumed that these safeguards are 100 per cent effective every time."

By the time January 28, 1965, arrived—the day for the Washington meeting with the industrial group—everyone had had a chance to go over the latest draft of the study, and everyone had his own ideas ready and waiting to nibble at it, if not actually to bite and snap. Cliff Beck opened the meeting at 10:20 A.M., thanking the industry men for pitching in to help with a difficult problem. While he didn't redefine the problem, its definition was easy to infer: How can you tell everybody in the country that an atomic plant could kill and destroy thousands of people, wipe out an enormous portion of the landscape and the Gross National Product, and expect them to like it? It was a challenge for the most suave public relations man in the world, and the committeemen were engineers and scientists, totally unskilled in the art.

The problem was particularly difficult because many of the industry men had hoped the situation might have improved since the old days of WASH-740. Instead the new figures showed that it was worse. Beck told the meeting that, although the accident experience to date had been good, there had been a number of "incidents" which had "been disturbing from the point of view of potentially serious accidents." He outlined the plan for the study: There would be two chapters, one on *probability,* handled by Beck and the AEC team. The second chapter would be the one on the *consequences* of an accident, handled by Ken Downes and the Brookhaven group. The latter chapter would not include all the details, and would more or less summarize the inescapable conclusion: "No inherent basis had been found on which to conclude that the consequences of a major accident would be less than those given in WASH-740, but could actually be greater."

This statement would be enough to make every insurance man in the country burn his actuarial tables. And as far as enticing the private insurance companies and thus taking the taxpayer off the hook, the government would have to continue to shield the utility companies from damage suits. But as Beck reminded the meeting, the AEC had promised the Joint Committee on Atomic Energy that a report would be written and produced. In other words, they were stuck with it.

Harold Vann, chairman of the industry group, admitted that there was a need to be frank, but he wanted to know more about the "model" accident Ken Downes had chosen. Downes explained that he picked the most likely event to cause an accident: the loss of coolant. Whether a reactor was a sodium-cooled breeder like Fermi, or a water-cooled pressurized reactor of the common run, the coolant liquid rushing through the fuel rods could never be lost without disaster—even in a portion of the core. With the loss of the coolant came the melting of both the fuel and its cladding, the rise to blistering temperatures, and the release of the lethal fission products to the environment.

He added that Brookhaven had chosen "not to get involved in the matter of probability, and had avoided talking in emotional terms, such as the number of deaths." (In the draft, he had avoided listing the awesome death figures that had been computed.) And he repeated the assertion that the possible size of the area of such a disaster might be equal to that of the state of Pennsylvania. There was no way the computations could cut down on these figures to any measurable degree.

Harold Vann's reaction was a natural one. He asked whether it would not be fair to conclude from these results that additional reactors should not be built.

Downes felt that Brookhaven was not saying that. Rather, it was saying that "complete reliance must be placed on engineered safeguards." Yet all the conversation at the meetings had reluctantly pointed to the fact they could *not* be relied on. In many cases, the "engineered safeguards" had created the accident, or made things worse. Further, the safeguards themselves could be destroyed in the initial accident, leaving the way wide open for a runaway. As Cliff Beck admitted later in the meeting, safeguards had been available in the SL-1 case at Idaho Falls, but the accident itself had inactivated them. A piece of machinery, turned into a missile by the start of an accident, could wipe out both the coolant system and the containment shell, as Beck noted.

Other industry representatives at the meeting noted that it appeared that Brookhaven had chosen the most unfavorable weather conditions for their accident model. But Brookhaven's Smith replied that the meteorological conditions they had used for the study "actually could have been somewhat worse." Forrest Western, an AEC safety man, said that it wasn't quite clear that the worst accident case was the one involving the worst weather conditions or the one involving the most people.

This brought up the question: Was there a real difference between an accident occurring in a country atomic power plant and one located in a city? Downes had already studied this question. He said that there wasn't too much difference in regard to what he called "the killing distance." He noted that for present-day sites, the wind could be assumed to head toward a city, and the radiation to fan out to cover a wide section of it when it reached there. If the power plant was in the city, its plume of radiation might cut a fairly narrow stripe through the city itself, fanning out beyond the city limits. It seemed that about the same amount of people would be killed either way. The narrow plume would hit a more concentrated population within a city; the wide plume would cover more territory with somewhat less intensity.

Cliff Beck reminded the meeting that they had to get the report out for the Joint Committee on Atomic Energy by early spring, since the Price-Anderson Act insurance hearings were coming up at that time. But the industry men balked. They felt that it would be better to wait for any "improved" results that might come from the continuing computations of the Planning Research Corporation probability studies. Marlin Remley, an industry representative from Atomics International pointed out that if the figures showed one chance in 500 reactor years for a catastrophe, and this were true, then the risk simply would not be acceptable. The question then would be: What *would* be acceptable? He even thought they should consult with the National Safety Council.

Dr. Merrill Eisenbud, a consultant from the NYU Medical Center, felt that it would be hard to present the picture even of small probabilities of an accident. He said that a low probability of leukemia due to fallout would be of more concern to a mother than the relatively high probability of an auto accident injuring the child. R. G. McAllister, a radiation specialist from the Liberty Mutual Insurance Company, brought up the question of human fallibility. The chances, for instance, of a ship blowing up at a Texas pier and killing over five hundred people would have been computed as very low. But it happened in Texas City in 1947 through human fallibility.

Forrest Western, a veteran investigator of the SL-1 accident at Idaho Falls, spoke up again to say that it was always the unknown that was difficult to judge, especially in a "rapidly changing field such as nuclear technology. One cannot foresee the rise of some unknown problem." He added that "there may be some limit to the risks that man may accept in terms of the size of the consequences even with a very small probability of occurrence. Nevertheless progress must continue, and one must face the question of whether to continue building more and larger reactors, even though the possible results of an accident might involve the area the size of Pennsylvania."

The meeting was now skirting the question of technology judgment versus the value judgment of the average man. Yet the average man would have no choice in the ultimate decisions. It was being decided for him. He was not only unable to act; he knew nothing about the apocalyptic casualty and destruction figures being talked about. Without this information, the ordinary citizen would have no motivation to act anyway. What was worse, as Marlin Remley brought up, if another study were done in 1970, the results would be even more frightening. Future reactors were being planned at twenty times the size of the reactors being discussed: 20 million kilowatts instead of 1 million.

At this point in the meeting Harold Vann, the industry chairman, suggested a lunch recess.

It was clear that neither the meeting nor the details revealed by the Brookhaven report were pleasing to the Atomic Industrial Forum. Harold Vann was very concerned by the overall conclusions drawn by the study. He doubted that large reactors would ever be built in view of the Brookhaven figures and the siting difficulties. The general opinion of the industry team was that all the work done by Brookhaven should be condensed into a few simple statements that merely said the results would be worse than those published in the WASH-740 version.

But Ken Downes felt that the full report had to be issued whether they wanted to or not. He said that too many people, including all five AEC commissioners, were familiar with the detailed effort underway at Brookhaven. While it wasn't stated at the meeting, this was the critical point about the whole affair. A definite report was expected, not only by the AEC commissioners and the Joint Committee on Atomic Energy—but by many private citizens and scientists who had learned about the planned publication from the congressional hearings that had prompted them. Everyone working on the study was caught in a vise. If an attempt was made to bury the report, or to deny it, all hell would break loose in the press and among the outside scientific fraternity.

The afternoon session of the meeting showed the results of the private caucus the industry members had held after lunch. They tried valiantly to knock down the pessimistic estimates that Ken Downes and his Brookhaven fellows had amassed. The death rate was challenged by Atomic Industrial Forum members, but Downes defended the figures he had gathered on the basis of a 1,200-rad dose that would kill fifty percent of the people exposed to it over a four-day period. This was a more than liberal allowance. Less than half this amount would do the same damage if the exposure came in one quick dose rather than spread over four days.

With this, the meeting ended. The new AEC-Brookhaven study posed a threat to industry that was so enormous that a chain reaction of desperation was bound to follow, one that would spread into the highest offices of the AEC. One thing seemed clear: The public must not see these appalling figures.

ELEVEN

At the Fermi reactor in Lagoona Beach, there was little time to think about the new Brookhaven study. Walter McCarthy was taking the lead in pushing through the infinitely complex details that would enable the reactor to generate up to 200,000 kilowatts, the next stage in moving toward the design target goal of twice that amount.

This was not an easy job. In addition to laboriously detailed reports on operations and safety, there were application hearings held by the AEC every step of the way. Since the Fermi reactor had been "brought to critical" in August of 1963, McCarthy was continuing to punch it through an elaborate series of tests to check out the design of the plant, and to iron out the many kinks that had slowed down the progress. "Bringing to critical" is simply like starting a car engine, and letting it idle. The reactor had been successfully started up over three hundred times; hundreds of exercises had been made involving the loading of the core, the behavior of the fuel, and the primary sodium cooling system on which so much depended.

Every setback that the Fermi reactor had suffered in its tests received McCarthy's personal attention. The bad check valves on the sodium pumps had been replaced with those of new design to prevent them from slamming, as they had during the tests. There had been slight, unexpected increases in reactivity—power surges not called for in the routine. McCarthy was sure he had this under control, and was well aware that a large increase in reactivity could mean a superprompt critical condition of great danger.

One mistake had taken place in loading a fuel subassembly, but no damage had occurred. McCarthy saw to it that the procedures were revised to take care of this sort of situation in the future. Radiation leaks seemed under control, as well as the single safety rod that had failed. The design of the "cask car"—that clumsy electric locomotive that carried the fuel—had been modified, and new instruments added. McCarthy was planning what was called a "power ascension program" which would take several months to creep toward the present goal of 200,000 kilowatts.

He realized that the hazards at this power level would be far larger than at the first experimental level. But he was convinced that the design of Fermi would allow no such thing as a sodium-water reaction within the vessel. The only type of "energy release" he felt it was necessary to worry about was an uncontrolled nuclear release that could only be caused by "a rapid reassembly of a critical configuration of nuclear fuel as a result of a gross meltdown of the core."

In presenting the case to the AEC to gain permission for the new rise in power, he pointed out that there were only three ways the Fermi core could possibly have a complete meltdown. One was a local melting in the core, which could spread to the rest of the core. Another was the loss of the sodium coolant from the primary system, with the fission product decay heating up the uncooled core. The third was a high power level that the coolant couldn't handle. This could start the sodium boiling, and a meltdown would follow.

"In my opinion," McCarthy told the AEC, "the design of the reactor makes any of these occurrences incredible." Then he added, somewhat contradictorily, "Let me state that I consider melting due to a local lack of cooling within a core assembly to be highly unlikely—but nevertheless a credible event." McCarthy was too much of a realist not to add the last qualification. But like those on the AEC and Brookhaven study committees, he was enamored of the phrase "highly unlikely."

While McCarthy was in the process of persuading the AEC to let him move ahead with the higher power level for the Fermi reactor, the industry members who had attended the volatile AEC-Brookhaven meeting at the end of January, 1965, were already beginning to let their displeasure be known concerning the public release of the report. Even before a final AEC decision to publish the report, industry was determined to quash it.

The resistance manifested itself softly at first. A small industry "working group" looked for every straw in the wind. They asked Brookhaven if some of the enormous fission products wouldn't "plate out"—be trapped—on the surfaces of the piping and containment vessel and thus cut down some of the damage to the public. But this had already been weighed and evaluated by Brookhaven, and had been found to make little, if any, difference in the catastrophic results.

The first sign of heavy pressure came in a letter from Atomic Industrial Forum's Harold Vann to Cliff Beck. In going over the drafts of the AEC study, he said: "This review has raised serious questions concerning the validity of the ground rules underlying the example accident analysis and, as a consequence, serious concern with the technical content of the draft report."

Vann questioned the whole purpose of the report. "To permit publication of the report without further exhaustive review," he wrote, "would, in our opinion, constitute a serious disservice to the [AEC] Commission, to the industry, and to the public. . . .

"If the Commission feels obligated to report to the Joint Committee prior to the Committee's consideration of the Price-Anderson [insurance act] extension, we recommend that the Commission forward to the [congressional] Committee an interim letter indicating that a re-assessment of the 1957 report is under way but is not yet complete. . . ."

Then Vann went on to recommend that the AEC initiate an entirely new and "much more comprehensive study."

Even if Beck had wanted to go along with these ideas, he was still faced with the threat of being accused of a whitewash or cover-up of the real facts that had come up in the study. No one liked the results, of course. Even Ken Downes didn't appear to be happy with the monstrous facts that had emerged from his deliberations.

But maybe Harold Vann did have one good idea. Perhaps the interim letter to the Joint Committee on Atomic Energy would smooth over the situation and buy time. It could be brief, and therefore not involve all the devastating numbers and predictions. It could suggest, rather than specify, the results.

Whatever Beck's reasoning, he decided to give Vann's interim letter idea a try. He sat down and drafted a letter to Congressman Chet Holifield, Chairman of the Joint Committee on Atomic Energy, on February 26, 1965. It read, in part:

> The Honorable Chet Holifield
> Chairman
> Joint Committee on Atomic Energy
> Congress of the United States
>
> Dear Mr. Holifield:
>
> During the Joint Committee hearings last year, you suggested that concurrently with the anticipated consideration of extending

> the Price-Anderson Indemnity Act, consideration should also be given to updating the 1957 report prepared by the Atomic Energy Commission with the assistance of Brookhaven National Laboratory on "Theoretical Possibilities and Consequences of Major Accidents in Large Nuclear Power Plants." This report is identified as WASH-740.
>
> Pursuant to your suggestion, we have asked the Brookhaven National Laboratory to re-evaluate its earlier study in light of today's technical information and current upward trends in power reactor sizes. It is the purpose of this letter to summarize the results of the laboratory re-evaluation, and to interpret the significance of these findings in the context of the 1957 calculations and in the context of present engineering practices and licensing procedures. . . .

Then, after a lengthy review of both the old and new studies, Beck continued:

> It is only in the highly improbable instance where these and all other engineered safeguards fail simultaneously that a loss of coolant accident could result in a public hazard. Calculations show that the upper limits in damages that could result from this hypothetical sequence of circumstances would not be less, and under some circumstances could be substantially more than the upper limits of the maximum consequence accident reported in the 1957 study.
>
> The higher inventory of fission products in the larger core would suggest that their unimpeded release to the atmosphere under the worst weather conditions, would result in higher radiation exposures and increased levels of contamination. This, however, is offset a little by the fact that recent experimental work has shown that a somewhat smaller fraction of fission products may be released from melting fuel than was assumed to be the case in the 1957 study. Another factor which makes difficult a precise comparison of the two sets of calculations is that the upper limits of radiation exposure and contamination in both instances are highly sensitive to unpredictable weather conditions.

> Much more important to the current re-assessment of the public hazards associated with power reactor operations than any refinement in our ability to calculate the damages that could result from a highly improbable maximum consequence accident is our increased understanding and appreciation of the engineered safeguards that have been developed to reduce both the probability and consequences of such an accident. . . .

After another long exposition of background on the studies, the letter continued:

> To sum up, we cannot categorically state that a power reactor will not experience a major accident even though we have had none during the eight years since the first power reactor went into operation. At this date, we have much more reason for believing that the likelihood of an accident occurring is even less than we believed it to be in 1957. Nor can we categorically state that one or more of the multiple engineered safeguards routinely incorporated in all power reactors will not at some time during the life of the reactor fail to meet its performance specifications. The probability of an accident occurring, however, before the malfunction of more than one safeguard has been detected or simultaneously with the failure of more than one essential safeguard appears to be very remote.
>
> Sincerely,
> C. K. Beck

After drafting the letter, Beck sent it to the AEC for approval, where it sat for over a month. Meantime he responded to Harold Vann in a letter that made it clear that Brookhaven had strictly followed the instruction to appraise the potential damages of the "upper limit" accident case. Then he continued with: "Of course the picture would be totally incomplete and inaccurate if the report contained only the theoretical calculations of consequence of possible accidents. Hence, as in the original WASH-740, we fully intend, if any report is issued at all, to have parallel, accompanying discussions of improbabilities of accidents, safeguards against accidents, contributions of safety research, a brief resume of the excellent record of power reactors, etc. But, as you know, this qualitative account will always appear weak beside the shocking results of the 'upper limit' case, if not in some of the lesser cases."

Beck let slip here a phrase that was most formidable. It was: "if any report is issued at all." If it were not to be issued, there would be as many repercussions as if the statistics were issued, and perhaps even more. He continued in the letter to Vann to say: "Your recommendations that we defer the report on the present study until we can report it in proper perspective on the basis of, and along with, a comprehensive study to find quantitative bases for probabilities of accidents, the dependability and effectiveness of safeguards in preventing accidents and limiting their consequences, etc., is one that presents us with a real problem. I don't mind leaving off any report of the present study indefinitely, but doubt the wisdom of committing ourselves to the study you recommend. I frankly have not thought of a way to go about it, who would do it, or what sort of outcome would emerge." He ended the letter: "Hopefully, we can find a simple way out of the dilemma, but this is not yet certain. I am continuing, therefore, to work with BNL on a more adequate version of a revised report."

By March of 1965, the American Public Power Association, a group of investor utilities, had joined in the fray, expressing "concern over the possible findings and subsequent public impact of the revision." One harried publicity man for the Pacific Gas and Electric Company told a meeting of the American Nuclear Society: "My task here, of course, is not to poke fun at the AEC, and I don't intend to do that. I have simply pointed out several areas which I believe they could consider improving in order to make it easier to get on with our chore of public understanding.

"One last suggestion along these lines would be to do us a

great big favor by cancelling the now-in-progress up-dating of the Brookhaven Report. . . . No amount of pointing to the disclaimers in the three-page covering letter in the front of the report could overcome those chilling words '3,400 killed, 43,000 injured, $7 billion in property damage' which rolled so glibly off their tongues from the depth of this theoretical [1957] study of the possible consequences of a major accident. . . ."

His apprehension, as a publicity man, was understandable. If he had trouble with the old figures, how could he possibly explain the 27,000 killed, 73,000 injured, and $17 billion damages of the new, unpublished report? Pacific Gas and Electric had been pointing out in lavish advertisements for years how its nuclear power plants were "good, clean and safe neighbors." One of their newspaper advertisements read: "The safety record of commercial nuclear power plants is unmatched in industrial history. . . . There have been no nuclear-caused deaths. Not even a significant injury." The same advertisement goes on to say that the reason homeowner policies have a nuclear exclusion clause is because of the Price-Anderson government insurance, neglecting to add that this insurance is paid by the public and would only cover a fraction of the potential damages from a major accident.

By the time St. Patrick's Day arrived, Beck was ready to take a vital step. He would lay the cards on the table for all five AEC commissioners. He summed up and reviewed the whole picture of the struggles of the special committees to come up with some kind of reassuring report. "It is an inescapable calculation," he wrote to the AEC commissioners, "that given the same hypothetical accidents as those considered in the original BNL study, damages would result possibly ten times as large as those calculated in the previous study. [Different estimates of damages during the study made these figures fluctuate.] Nothing has been discovered in the interim which assures that such accidents cannot happen, even though we are convinced that the probability of such accidents is believed to be lower than the low estimates of the likelihood of such accidents made at the time of the original study."

He then informed the commissioners that the Atomic Industrial Forum had strongly urged that the new Brookhaven study not be published in any form at the present time, and that a letter, such as the one he had drafted, be sent to the congressional committee instead.

On March 31, 1965, the Commission approved a watered-down letter, and a watered-down version of the new Brookhaven report, to be called an "unclassified version." All the terrifying numbers of deaths, injuries, and property damage were missing. None of the details of damage to infants, adults, and crops was specified. All the report amounted to was that the writers had to come to the "inescapable conclusion" that the theoretically calculated damages would "not be less, and under some circumstances would be substantially more, than the consequences reported in the earlier study."

There were vague promises that "later reports" would reveal the methods of calculation for this conclusion. The official commitment made for a thorough study and updating of WASH-740 was tossed aside.

Stan Szawlewicz, a manifest realist, sensed clearly the problems that would arise from the watered-down version. In his Official Use Only comments on the new version, he said: "The length of the discourse suggests that a lot more has been done than what appears in print and will inevitably lead to key questions as to just what were the consequences of the Brookhaven study. . . . The real unpublished results of the BNL study shows very simply that the consequences (number of people killed) is directly proportional to the increased fission product inventory of the larger reactor. . . . In practical terms, it is difficult to assign a meaningful probability number to such an accident and little consolation can be derived from the statement that it should be very small. . . ."

He went on to make many suggestions for a more positive approach in a different kind of report for the public that would be more reassuring. His final comment showed this concern: "In a general way, the increase in reactor size and fission product inventory appears to be the major difference between the old and new hypothetical accident calculations with improved treatment of meteorology. This suggests that future similar analyses of even larger reactor sizes . . . can give even more catastrophic results. If we emphasize the 'potential' of such reactors to do damage, as is the case with the hypothetical accident studies, we doubt if they will ever be built."

The eventual letter to Chet Holifield, signed by Glenn Seaborg, AEC chairman, was so scraped to the bone that it said practically nothing except that the likelihood of major accidents "is still more remote," the "consequences could be greater," and that the Price-Anderson Act insurance was needed more than ever, and should be extended.

Although the report that the Joint Committee was promised by the AEC had somehow vanished into thin air, the congressional committee held the insurance hearings on the Price-Anderson Act in June of 1965, anyway. There was no new evaluation to go on. The minutes of the AEC-Brookhaven meetings were safely locked behind closed doors, unavailable to the congressmen who had asked for them.

It seemed obvious to some observers that, if the estimated destruction figures had been brighter because of better engineered safeguards, the insurance companies and the utilities would have had confidence to take on the insurance burden. But all the utilities that testified at the hearing stated flatly that they would stop building any atomic power plant under construction and never plan another one unless the government took on the mammoth responsibility of protecting the public, and relieving them of any damage suits. It was the government indemnity supplied by the taxpayer—or else. In spite of the appalling potential damage figures, still hidden from the public, the Price-Anderson insurance act was extended by Congress without protest.

In a sort of a farewell memo to the steering committee, Cliff Beck said that the final handling of the "revised Brookhaven report" has been made up of "a long series of complicated maneuvers and negotiations during efforts to arrive at a final written document which would be acceptable to all responsible parties. During these efforts it simply was not possible to accomplish the many discussions, drafts and re-drafts through the awkward mechanism of a widely scattered steering committee."

Beck then thanked the members for their assistance, and finally confessed that the "revised Brookhaven report" had not been completed. Thus, by the end of June, 1965, the whole matter seemed to be resting peacefully in its grave.

It did so, in fact, until the middle of August, a year after the mess had started. At this time, John Palfrey, the AEC commissioner who had begged off talking about the new Brookhaven report in his California speech, received a short letter from David Pesonen. Pesonen was a young lawyer intent on protecting the environment in Bodega Bay, near San Francisco. He wrote:

August 14, 1965

Dear Commissioner Palfrey:

In your address to the meeting of the Atomic Industrial Forum in San Francisco last December, you mentioned that the Brookhaven Report (WASH-740) was in process of being updated and would be ready shortly.

So far I have seen no notice that the report has been completed and made available.

Can you advise as to its current status. If it is completed, how may one obtain a copy?

Commissioner Palfrey wrote back on August 27. He said:

Dear Mr. Pesonen:

The results of the re-examination of the Brookhaven Report were described in a letter from Dr. Seaborg to Chairman Holifield, and this was entered into the record of the hearings of the Price-Anderson. I enclose a copy of that letter for your information.

Pesonen was not satisfied. The letter to Holifield was far from a report or a study. It was two pages filled with vague generalities. He replied to the commissioner on September 13, 1965:

September 13, 1965

Mr. John G. Palfrey, Commissioner
U.S. Atomic Energy Commission
Washington, D.C. 20545

Dear Commissioner Palfrey:

Thank you for your letter of August 27th, forwarding a copy of Chairman Seaborg's letter of June 18th to Congressman Holifield, concerning government indemnity for power reactors and the scheduled up-dating of the Brookhaven Report, WASH-740.

However, my letter was concerned with the current status of the Brookhaven up-dating, which you mentioned in your address on Price-Anderson at the AIF meeting in San Francisco last December. Specifically, you said that the Brookhaven report's revision "is not ready yet," and I was interested in when it will be ready.

Chairman Seaborg's letter suggests that such a study has been completed when he states that "a restudy of the theoretical consequences of hypothetical accidents by our staff and that of Brookhaven has led us to fairly predictable conclusions."

I get the impression that your letter to me is not entirely responsive to my original request. When will the "restudy" mentioned by Chairman Seaborg be available for public inspection?

Sincerely,
David E. Pesonen

Commissioner Palfrey's reply stated flatly that no new report was "in existence or contemplated." However, Palfrey modified his stance somewhat by saying that two "technical reports" on meteorology and health physics would be coming out of Brookhaven "for public distribution." There was no mention of Brookhaven's major effort—the detailed impact of a theoretical power plant accident on the population and the land.

Pesonen had no idea of the appalling figures that had emerged from the restudy, but it was becoming obvious that something was being covered up by the AEC, and he was determined to get at it.

Pesonen also wrote occasional articles for magazines on a freelance basis, and he began digging. He found that the Price-Anderson Act was the only way the utilities could be protected from damage suits resulting from death, injury, and destruction from a nuclear accident. So Pesonen got busy, and wrote an article he submitted to *The Nation* called "Atomic Insurance: The Ticklish Statistics." It appeared on October 18, 1965. In it, Pesonen suggested that the AEC was suppressing a new report concerning the consequences of a major accident in a large nuclear power plant. Pesonen didn't know this for sure, but he had guessed enough from Commissioner Palfrey's evasiveness to suggest it.

The article jolted the PR men in the AEC's division of public information. Duncan Clark, director of the AEC's press problems, lost no time in developing a ready-made "boiler-plate" response for the entire board of commissioners, who would undoubtedly be faced with some embarrassing questions as the result of the article. The stock answer Duncan Clark prepared for the commissioners was this:

Q. David Pesonen has suggested in an article in *The Nation* magazine that the A.E.C. has suppressed a report which deals with the theoretical consequences of a major accident in a large nuclear power plant. He refers to this report as an updating of the 1957 Brookhaven report. Is this so?

A. In connection with Congressional consideration of an extension of the Price-Anderson indemnity law for an additional 10 years to August 1, 1977, members of the A.E.C. staff and the staff of Brookhaven National Laboratory reviewed the 1957 study on theoretical consequences of a major accident in a large nuclear power plant (known as the Brookhaven report).

While this review was going on, no one knew what the results would be or the form they would take, but the process was commonly referred to as the "updating of the Brookhaven report."

No new report is in existence or contemplated. It was the judgment of the persons from Brookhaven and the A.E.C. staff at the conclusion of their review that no detailed refiguring of the 1957 report was needed to provide the Congressional Joint Committee on Atomic Energy with the information it needed to consider extension of Price-Anderson indemnity.

For someone familiar with the elaborate goings-on at the frequent AEC-Brookhaven meetings, the publicity department's reply might seem ludicrous. But to the average outsider, it sounded quite plausible and disarming.

The boiler-plate copy was prepared just in time, because inquiries began coming in promptly in response to the article. On November 8, 1965, Senator Abe Ribicoff sent along to the AEC a letter from one of his Connecticut constituents. He accompanied it with a brisk note: "I would appreciate a full report on the matter raised in Miss Taylor's letter."

Miss Taylor was a citizen of Glastonbury, Connecticut, and was concerned. When she read Pesonen's article in *The Nation*, she found it provoked many questions in her mind. "If the

Atomic Energy Commission is in fact suppressing the consequences of a major accident in a large nuclear power plant, this is certainly cause for public concern," she said in her letter to the senator. "Moreover, since privately-owned public utility corporations are in effect guaranteed a reasonable profit by the mode in which their rates are set, why should they be pressing for the development of a source of power which carries with it the possibility of disaster? I hope you will be able to give this matter some attention, as it is certainly a situation which needs to be investigated and exposed for public consideration."

Harold Price, the AEC director of regulation, answered Ribicoff with a letter that neatly incorporated the language of Duncan Clark's carefully prepared statement:

Nov. 29, 1965

Dear Senator Ribicoff:

This is in response to your referral of November 8 requesting information on questions arising from publication of an article in the October 18, 1965, issue of *The Nation* entitled, "Atomic Insurance: The Ticklish Statistics." This article was cited by Miss Patricia Taylor in her letter to you of October 24.

The article by Mr. David Pesonen suggests that the Atomic Energy Commission may have suppressed a "report" dealing with the theoretical consequences of a major accident in a large nuclear power plant. The "report" referred to by Mr. Pesonen is described in the article as an updating of the 1957 report, "Theoretical Consequences of a Major Accident in a Large Nuclear Power Plant." This 1957 report, made public at the time, was prepared at the Commission's request and was submitted to the Congressional Joint Committee on Atomic Energy in connection with its consideration of proposed legislation which ultimately resulted in enactment of the Price-Anderson Act in 1957. The report is commonly referred to as the "Brookhaven Report," since it was prepared primarily by a group of scientists from the Commission's Brookhaven National Laboratory in New York.

> In connection with the recent Congressional extension of the Price-Anderson Act, members of the AEC staff and the staff of the Brookhaven National Laboratory reviewed the 1957 report. The reviewers determined at the conclusion of their study that no detailed refiguring was required to provide the Joint Committee on Atomic Energy with the information it needed to consider extension of the indemnification legislation. The conclusions of their review were set forth by Chairman Seaborg in his letter of June 18, 1965, to the Joint Committee (copy of which is enclosed), which was made public. A copy of a letter from Commissioner Palfrey to Mr. Pesonen, dated October 8, 1965, also is enclosed, which gives further information on the review. . . .

In the brief history of atomic power plant development, there had now been three thorough and reasonably objective studies made of the possible consequences of a major power plant accident. Each one of them had brought forth figures of doomsday proportions. In the maximum cases, WASH-740 predicted 3,400 dead with the smaller reactors of the 1950's. The University of Michigan study under Professor Henry J. Gomberg had shown a potential for 133,000 dead from the Fermi reactor at Lagoona Beach. The WASH-740 update had come up with 27,000 dead, although the AEC was pretending that this update didn't exist.

Somehow, the public relations men at the AEC would have to get around the dilemma, because some kind of study had been announced. A few solutions were gradually emerging. One was the launching of a long and handsome report called WASH-1250; its title was eventually to become "The Safety of Nuclear Power Reactors." It was a laundered version of the restudy, presenting none of the destruction and casualty figures.

Another was the AEC's launching, a few years later, of a $3 million probability study of reactor accidents, under the direction of Professor Norman Rasmussen of MIT. The decision to attempt this after the failure of the earlier probability study was based on the hope that somehow the wizardry of statistics could be shaped to come up with a more favorable picture. Nonetheless, there were many on the Brookhaven and some on the AEC committee who were convinced this could not be done. As Clifford Beck wrote in his appraisal of the attempt to figure the odds on a major accident: "Here is encountered the most baffling and insoluble enigma existing in our technology: it is in principle easy and straightforward to calculate potential damages that might be realized under such postulated accident conditions; there is *not even in principle an objective and quantitative method of calculating probability or improbability of accidents or the likelihood that potential hazards will or will not be realized.*" [Italics added.]

Professor Rasmussen would be facing stiff examination when he eventually got around to publishing his figures nine years later.

TWELVE

At the Fermi reactor at Lagoona Beach, things looked bright at the start of the new year of 1966, in spite of the chilling January winds whipping in from Lake Erie. The overriding excitement of the operating tests, which were just beginning to push toward the point where electric power could be produced, was tempered by the exacting routines that governed safety.

The step-up operation would inch with meticulous care and caution toward the first goal of 200,000 kilowatts of thermal power in eight carefully plotted steps that would take up the large part of a year. At the start, the reactor would barely peak up above the 1,000-kilowatt level. It would creep toward 20,000 kilowatts by the fourth step, and at the sixth step, it would move halfway to the goal. As the tests passed the 80,000-kilowatt level, it would be able to pump out some token electricity to Detroit Edison's customers. This would be a landmark day, after the years of frustration and delays and a cost that had now reached well over $100 million.

There was drama in the invisible storm of neutrons that the fuel rods sprayed at each other as they shot out with blinding speed. The operator at the control board could only sense this power through his instruments and dials. But as the control rods slowly withdrew, and the instrument readings reflected this silent power when the huge pumps sent the sodium syrup through the system, vibrations were felt in the floor of the control room that hinted at the reactor's awesome power.

There were, of course, all kinds of automatic safety devices. When the safety rods were withdrawn after reloading the Fermi reactor, they automatically stopped at two different check positions so that the count rate could be checked before proceeding further. The only way the process could continue was by a single key in the possession of the shift supervisor. If everything checked out at that point, he could override the interlock only by inserting the key and turning it. Otherwise, it could not proceed. Most important, and what would be watched constantly, was to make sure that the coolant flowed properly at all times. The design of Fermi was such that if the coolant should be lost because of a pump failure, the reactor would scram itself automatically, and auxiliary pony motors would take over immediately. These precautions were necessary as the boiling point of sodium could never be reached without disaster.

Even though it was constantly disclaimed, a nuclear explosion could occur if a fast breeder reactor like Fermi was brought to "superprompt critical." The final studies on the EBR-I meltdown had shown that if the shutdown attempt had taken place one-half second later than it did, the reactor would have exploded. Some computer models of breeder accidents showed that if fuel damage and melting took place, and if some of the coolant were shoved out of the tiny channels between the fuel rods and assemblies, the fuel could be jammed and compacted together to turn the neutrons into a hornet's nest of activity. Then a small explosion might occur that would push the fuel rods even closer together. The closer the fuel rods are packed, the greater the unwanted power surge. After that: a larger explosion adding to the first. In other words, a small unplanned power surge inserted into the core would lead to a small explosion, feeding the reactor with a large power surge, followed by a large explosion.

Just as burning logs fall and set others on fire, so could the fuel assemblies crash and fall, leaving unpredictable gaps in the once tidy core and blocking other coolant channels. A rule of thumb had been established that the size of an "energy release" is proportionately larger in a big core than in a small one. Comparing the Fermi reactor with the midget EBR-I, an explosion in the Fermi core could be up to fifty times the designed limit of its containment shell. But the carefully studied Hazard Report filed with the AEC for the Fermi reactor dismissed this idea as "incredible."

Another dangerous threat would be if the fuel melted and mixed with the sodium coolant. This in turn could cause a sodium vapor explosion that could be even more violent than a nuclear burst. It was obvious that there was little, if any, margin for error.

But the crew that had worked so long and so patiently was forewarned and forearmed about these types of problems. They worked with confidence during the meticulous process of edging the power up slowly, week after week, toward the 80,000-kilowatt level. The men putting the final touches on the reactor, before it would actually produce electricity for the first time, were a dedicated lot.

There was the swashbuckling Walter McCarthy. Tall and wiry, he moved about with an air of brash confidence and restrained tension. He displayed a dry Irish wit that seemed to soften his impatience with any form of bungling. His passion for detail was enormous.

Like Cisler, he was a Cornell man, but, in contrast to Cisler, he was freewheeling and hyperthyroid in his movements and action. Cisler's body language was frozen and sphinx-like.

McCarthy's was explosive, abrupt, and dynamic. Cisler, who spent a considerable amount of time at the Fermi site, would think and examine and reflect with an air of pontifical restraint. At one test he stayed until nearly four in the morning to observe and counsel.

Wayne Jens, a technical assistant for Fermi, was in marked contrast to McCarthy. A graduate of Purdue and the University of Wisconsin, he was reflective and incisive in his thinking. He and McCarthy complemented each other; they worked together smoothly.

Eldon Alexanderson, assistant superintendent and reactor engineer, matched Cisler's quality of restraint and soft speech, but somewhat as if there was a fire inside he wanted to let out. He was meticulous, precise, somewhat of a worrier. Under Bill Olson, the on-the-line chief of the operation, he was responsible for an infinite number of details as they pushed toward the milestone target of 80,000 kilowatts—the point where the reactor would be able for the first time to send a portion of its electrical power over the grid on the Detroit Edison system.

These firing-line scientists looked on their work as a magnificent challenge that they could not only handle, but conquer. The plans, specifications, blueprints, invoices, bills of lading, and engineering reports made up a nightmarish jigsaw puzzle that would dismay all but the lionhearted.

Even such a thing as the operating license was a maze of complexities. One paragraph read that the Fermi reactor was now licensed ". . . to receive, possess and use 41,900 kilograms of depleted Uranium-238 contained in radial blanket subassemblies and the axial blanket sections of core subassemblies, 17,500 pounds of depleted Uranium-238 constituting shielding for the cask car, 14,245 pounds of depleted Uranium-238 contained in derbies, blanket rods, simulated fuel pins, and solid specimens, 250 grams of Thorium-232 as foils or other shapes suitable for neutron spectrum measurements, 0.1 gram of Uranium-234 and

0.1 gram of Uranium-236 in fission counters, pursuant to Act and Title 10, CFR, Chapter 1, Part 40, 'Licensing of Source Material.' "

Hundreds and hundreds of specifications like this had flowed through the process of putting this giant Swiss watch of a reactor together. And through it all, *there could be no mistake.* What if 0.1 gram of Uranium-236 got thrown out with the packing carton? Stranger things than that had happened, and gone unnoticed. One by one, McCarthy, Jens, Alexanderson, and the team checked, double checked, triple checked every item with the passion of a tent-meeting, pulpit-pounding pastor. Could there be any leakage of the sodium coolant? Were the battery-powered pony motors ready to pump the sodium automatically on the loss of normal electric power? Would the automatic scram work as designed? And most of all, was the "maximum credible accident" that was covered so thoroughly in McCarthy's Hazard Summary Report really the worst that could happen? In McCarthy's opinion, the worst was covered as his hazard summary read: "The melting of some or all of the fuel in one core subassembly, due to either complete or partial plugging of the nozzle of that subassembly. . . . I have described the mechanics of meltdown and the basis for my opinion that if fuel melting occurs, it will be confined to the subassembly in which it happens."

The nozzle was a snout at the bottom of the tall, slim stainless steel wrapper can of curtain-rod-like fuel pins. The gooey sodium syrup would rush up through the thin spaces between the pins, and keep them cool and tame. At the same time, it would convey the heat away from them, as it rushed over to the steam generator building through closed pipes to indirectly create steam. The pipes would be hot enough to boil the water that surrounded them in the heat exchanger, without their sodium contents ever coming directly into fatal contact with the water.

If the coolant flow was ever blocked, McCarthy's computations figured that the meltdown would not spread from the single plugged-up subassembly. This was very important. If it spread to others, there would be hell to pay. Some scientists were sure that if the melting spread to other subassemblies, the results could lead to disaster, as the molten, waxy uranium fell down through the core. Here the question of the mild-sounding but terrifying condition known as "secondary criticality" would have to be faced.

The zirconium-plated, inverted ice-cream cone at the bottom of the coffee-pot-like reactor would play a big part here. If some of the hot thick molasses-like uranium substance were poured over the inverted cone, the chances were it would spread out thinly into a shape that would probably not go into a critical stage. But, like a ton of thick mud dropping down through a jungle gym, it was hard to tell exactly what shape might form. If a thick glob of it landed in a hunk and froze at that position, the slightest disturbance could set off an explosion, or, as the engineers preferred to call it, an "energy release." What could be done if this happened inside the sealed-off reactor, so drenched with the opaque liquid sodium that no one could tell what shape the molten fuel was in?

There would be days, perhaps weeks, of tense, anxious suspense, as readings and procedures were worked out. The crew probing an accident would have to walk on eggs as they tried desperately to figure out what they could do about it. There could be no "reactor-nauts"—the term they used for the space-clad mechanics—sent down into the bowels of the reactor vessel now. It would be so radioactive that no one could survive in it, regardless of masks and protective clothing. During the previous sodium leak, the crew had been able to get down inside the viscera of the reactor because there had been only the sodium to worry about. That was before the fuel had been put in, and there was no radioactivity. The danger to the crew then had been bad enough; the dangers now would be unthinkable.

But McCarthy's careful analysis did not neglect the possibility of "secondary criticality"—a condition that could be far worse than the initial accident. He was most confident that such a thing would not result from a meltdown. However, just to be on the safe side, he had assumed in his analysis that it would happen. The silent splitting of quadrillions of neutrons would increase rapidly in three seconds. Then they would spit even faster as the melting increased. The faster this happened, the greater the "energy burst." McCarthy, along with Hans Bethe, was sure this would not be more than the equivalent of five hundred pounds of TNT, which the containment shell was designed to hold.

It was this 500-pound TNT estimate that was a special bone of contention in some quarters. Some critics claimed that the Fermi Hazard Report estimates were much too low, and that they had been set simply because this was the limit that the containment shell could be designed for without the cost soaring impractically high. If an "energy burst" zoomed up to fifty times the Fermi estimate, which some critics thought possible, all the neat figures and computations would be hellishly violated at the expense of Detroit, the state of Michigan, and all the ships on Lake Erie. P. M. Murphy, a General Electric nuclear energy executive was to say a few years later: "It is, in our view, unlikely that one will be able to design for the worst accident permitted by the laws of nature, and end up with an economically interesting system, even after additional research and development has been carried out."

While the painstaking tests were going on at the Fermi plant during the first few months of 1966, the last spasms of the ill-starred Brookhaven report update were taking place. Even the Advisory Committee on Reactor Safeguards could not get access to the complete Brookhaven files, since it was not an official part of the AEC. Although this committee was the only thing that resembled a scrawny watchdog in the entire AEC oligarchy, it had been blatantly ignored before, back in the fifties. (This was when Lewis Strauss had buried the Advisory Committee's letter expressing alarm and concern about the Fermi plans.) Therefore, the hiding of more bad news from the Advisory Committee was not unexpected.

Cliff Beck attended one of the Advisory Committee meetings in the late spring of 1966. He made it plain to the safety group that the techniques used by Brookhaven needed "much more editing before publication is possible." He also added specifically and rather defensively that he didn't want "the conclusion written down." But the Advisory Committee was not satisfied with this. They wanted the exact conclusions, and they wanted to see them in writing.

By now, Ken Downes and his Brookhaven colleagues had reached the conclusion that "if commitments on the publication of such a report had not been publicly made, they probably would choose not to complete the report." When this was brought up at the Advisory Committee meeting, temperatures began rising and frustration began bulging at the seams. The result was that the Advisory Committee on Reactor Safeguards told the AEC men to direct Brookhaven to compile the conclusions and come up with the significant findings that had been worked over in the laborious WASH-740 restudy.

If Brookhaven was worried about disclosing this, in the light of all the pressure to squash the report out of existence, the Advisory Committee suggested that the figures and conclusions could be supplied in an informal, unsigned memo, so that the source would be unidentified. It could be completely anonymous. Both Brookhaven and Cliff Beck protested that this would require too many man hours, although there is no evidence that they told the Advisory Committee about the endless drafts of the restudy that were hidden away.

The Advisory Committee refused to be stalled, however. They said that if they didn't get the straight, amassed facts, it might be necessary for them to write a letter to the AEC commissioners about the situation.

There was flak everywhere as a result of this conflict. It was followed by massive attempts at stalling. The Advisory Committee stepped out of the usual tangle of bureaucratic confusion and persuaded Brookhaven's Winsche to "volunteer" to appear before them at the Argonne Laboratories, near Chicago, and bring along with him "some written pages."

When Cliff Beck heard about this, he was upset. He didn't want anything left around in writing at all. He asked the Advisory Committee to wait, but they flatly refused. They said that if they couldn't have written material, they at least wanted information that could be written on a blackboard, so that the Advisory Committee could copy it down.

When Beck told Dave Okrent, who was now on the Advisory Committee, that he would have to be away during the time for a suggested meeting, Okrent told Beck that this would be great, because then the AEC men would not officially know what information had been given out at the meeting, and therefore Beck would be off the hook.

But in spite of all the maneuvering behind the scenes, the real results of the Brookhaven restudy of the old WASH-740 report were kept out of sight. Instead the AEC publicity men went to work to shape what they hoped would be convincing reasons why the promised report would not be forthcoming. Elaborate phrases and stock answers continued to be supplied to all the AEC commissioners, so that they could answer any queries from the public, the Congress, or the press. Some statements were: "The Commission has no plans to update the report WASH-740," or, "There have accumulated more than 780-reactor years without a single radiation fatality or serious radiation exposure." The men who, at great personal risk, had dragged out two bodies from the SL-1 reactor, and scraped the third one from the ceiling into a net might have gone through some head-scratching when they read the latter statement, not to mention those who risked their lives to prevent disaster at Windscale and Chalk River.

In the meantime, the publicity men welcomed people like Philip Abelson, editor of the prestigious *Science* magazine, who wrote a glowing piece in May of 1966. "We are now in a new era," Abelson's editorial went. "Atomic energy has proved relatively safe, reliable, and clean. Radioactive wastes can be contained. The bookkeeping cost of nuclear power has become competitive. . . . A continuing effort is being made to guarantee reactor safety. . . . In the contest between conventional and nuclear power, the balance is shifting rapidly. In a few years most new major planned installations are likely to be nuclear."

The AEC publicity men were not happy with responses like the one by Sheldon Novick of Washington University, that soon appeared in the letters-to-the-editor column of the same publication. Abelson, however, was gentleman enough to print it in his own publication.

"Abelson repeats the often-heard comment that electric power will be increasingly produced in the country by nuclear reactors," Sheldon Novick wrote. "In view of the many unfavorable aspects of nuclear energy, the scarcity of Uranium-235, the enormous expense of reactor development, and the inherent pollution and explosion hazards, it is puzzling that the efforts of scientists and the resources of the United States have not been directed toward developing a less hazardous energy source. Still less comprehensible is the pressure to proceed with construction of reactors in a full-scale commercial program before the dangers involved are better understood. After all, we have a century or so to find a replacement for fossil fuels."

The last sentence was important. With many centuries of coal available, why the rush into a catastrophe and disaster that was entirely possible, if not probable? Coal, if it did nothing else, would buy time—especially on a crash liquefaction and gasification program. That would reduce pollution problems. Land damage? Yes, but land restoration in some areas was possible at a fraction of the cost of a single nuclear runaway.

Letters like Novick's, however, did nothing to dim the enthusiasm of the Fermi crews as they continued their tests, and climbed up the ladder to what they were sure would be certain success. Thanks to the Joint Committee of Congress and the taxpayer-financed Price-Anderson Act, there was nothing to worry about now in the way of insurance. Even if the impossible happened, and the 133,000 estimated by the University of Michigan study were killed by a major accident at Fermi, the reactor corporation would be totally exempt from any claims brought against it for death or injury or property damage. In this hypothetical case, the claims could run anywhere from $10 billion to $30 billion, and the utility company could continue in business as usual and not have to worry about its assets being touched. But the Price-Anderson law provided an absolute ceiling of $560 million in accident insurance. It simply put a lid on the amount that would be paid out to the sufferers that happened to be unfortunate enough to live in Michigan or northern Ohio.

Under the Michigan University theoretical accident toll of 133,000, death claims alone from the accident could run to over $7 billion, if a miserly $50,000 per death was allowed by the courts. But the total amount to be paid out to the victims would be the $495 million government portion under the Price-Anderson law restrictions, plus $65 million offered by insurance companies. This would divide into about $4,000 per life lost. But the injured and property damage would also have to receive a share. If the Brookhaven estimated property destruction of $17 billion were added to this, the total damages could come to over $24 billion. The Price-Anderson provision would leave a gap of over $23 billion. Congress would undoubtedly try to come to the rescue with some kind of emergency aid, but how could the taxpapers be able to gather that much money in addition to their own tax burdens?

The only thing to do was think positively and pray that nothing would happen—and to make sure that the Fermi reactor and its crew were infallible. There was still that undeniable inspiration to reach for an historical engineering first: the first breeder reactor to pump out electrical power over commercial transmission lines. It would be a soothing poultice for the Fermi engineers after the long stretch of tribulations.

August 6, 1966, was the day that it finally came about. Starting that Friday afternoon, and for fifty-two hours, the scorching hot sodium churned through the pipes and created enough steam at 100,000 kilowatts of heat to produce 33,000 kilowatts of electric power. Ironically, however, this was less than half of what the standby oil generator at the Fermi site could turn out. It was, however, as Walker Cisler proudly said, "the highest power level yet reached by a breeder reactor plant."

But the triumph was short-lived. Not only had the costs now mounted to $120 million over a span of a decade of problems, but setbacks still plagued the project. There had been troubles showing all along during the 1966 test program with the thermocouples—those "oven thermometers" that were sprinkled among the fuel subassemblies to make sure the fuel was staying in line as far as fuel temperature was concerned.

One of the subassemblies, known as M-091, had been particularly temperamental all through the tests. It was hard to tell from the readings whether it was too hot, or the device was not registering correctly. Subassembly M-140 was showing similar signs of discontent. So were several others. Since any hot spot in a single subassembly could be a harbinger for disaster, these anomalies were watched and checked very carefully. And there were more problems with safety rod number 3. Steam generator leak repairs joined with all the other problems to keep the Fermi reactor out of any major action until October rolled around.

In fact during the brief moments of triumph on August 6 and 7, three subassemblies showed abnormal temperatures. It was decided to shift their positions around in the egg-crate-like structure that held the core together. In this way, an added check could be made to see if the fault was with the instruments or whether the subassemblies were actually abnormal. Actually, the high temperature readings were strongly suspected of being false, because all the other subassemblies around them were reading comfortably normal.

By October 4, 1966, the Fermi engineers had things in good enough shape to make another try at reaching their first-stage high-power goal. They planned to run the reactor for a while at idling speed, slowly raising the temperature of the viscous sodium fluid to a little over 500°F., a temperature high enough to make pressurized water boil, but not sodium. There were three routine tests to make, mainly checking pressures and temperatures. At eight o'clock at night on October 4, the system was ready to make the approach to criticality with the slow withdrawal of the control rods. It took until 11:08 P.M. for criticality to be reached, that point at which the reactor would stand by for the next step.

Here, the process was stopped for the control room operators to check everything out. The readings on the instruments were exactly as predicted for the amount of distance the control rods had been withdrawn. This is an important check, because the higher the long, thin control rods are pulled out of the core, the greater the power that should be coming through. If the rods are out some distance and the power is less than predicted, it is an immediate sign of trouble.

There was no trouble apparent at this time, however. Things looked good for the big push which was to begin at 8 A.M., October 5. The Fermi crew was naturally itchy to get on with the job after so many ceaseless delays.

The first problem that was discovered on the morning of the fifth was a malfunction in one of the steam generator valves. It took until nearly 2 P.M. to clear it up. Then another power push was made. But this was barely started before there was trouble with the boiler feed-water pump.

Again the control rods were pushed back in to reduce the power while this was taken care of. After another start-up, there was a brief hold to put the reactor on automatic control. Then the power began rising again. By 3 P.M., the power was up to about twenty percent of its 100,000-kilowatt limit in the current series of tests.

It was at this point that Mike Wilber, the assistant nuclear engineer in the control room, noticed some erratic changes in the neutron activity of the reactor. However, this situation had been noticed before at about the same power level. It had been thought to be a pickup of electronic noise in the control system. The control system had not been affected when this had happened in the past. But just to be on the safe side, the reactor was put on manual control again, and the reactor behavior watched carefully.

In a few moments the apparent noise disappeared from the instrument readings. Again, the reactor was put on automatic control. Any decision at the control board now would be critical.

THIRTEEN

Just a few minutes after the first signs of ill behavior at the control panel of the Fermi reactor, at 3:05 P.M. to be exact, Mike Wilber noticed another problem. For the amount of heat and power that was coming out of the reactor, the control rods should have been raised only six inches out of the core. Instead, they were a full nine inches out. This was not a comfortable situation. Further, the reactivity signal was again moving crazily and Wilber's first thought was that the core temperature was too high.

The instruments that showed the temperatures of the individual subassemblies were rather awkwardly installed, about twenty or thirty feet away from the main control board, behind the relay panel—a wide bank of instruments stretching along the width of the control room.

The operator stopped the power increase immediately, and Wilber went behind the control board to check the core outlet temperature instruments. He scanned them quickly. It was immediately obvious that two subassemblies were showing high outlet temperatures: M-140 and M-098. Each tall, slim can that wrapped a bundle of slender fuel pins had its own designation in the core, just as a crossword puzzle has its squares identified. M-140 has been acting up before. It still wasn't clear whether it was the instrument that was off, or whether the subassembly itself was actually overheating. The instrument had been reshuffled to a new position to check this, because of the previous misbehavior. But M-098 had never been a problem child. And it had never been moved from its original position.

It was hard to get a complete picture of the blistering hot core, because only one out of every four subassemblies was equipped with a thermocouple. A most disturbing thing was that M-140 should be reading about 580°. It was now showing over 700.

Suddenly, as Wilber was standing in front of the temperature instruments behind the control panel, radiation alarms went off. It was exactly 3:09 P.M. The air horn began blasting—two blasts every three seconds. Then, over the intercom, a laconic announcement: "Now hear this. Now hear this. The containment building and the fission product detector building have been secured. There are high radiation readings, and they are sealed off. Do not attempt to enter. Stay out. Both buildings are isolated. This is a Class I emergency. Stand by for further instructions. Stand by for further instructions."

The crew began scrambling about on its assigned emergency procedures. All doors and windows were closed. The fresh air intakes in all buildings were shut down. Plant guards closed off the entrances to the site. The health physics team rushed to the control room. Someone—no one remembers who—phoned the local Monroe County Sheriff's office and the state police headquarters at Lansing.

A Class I emergency was in effect at Lagoona Beach, and no one could say what would happen next.

Regardless of how well trained and prepared the Fermi crew was to meet emergencies, the situation in the control room was tense and dangerous. A sudden appearance of radiation, of poisonous fission products leaking out of the reactor vessel and into the containment building needed immediate expert attention and extremely cautious action. There were plenty of words covering the situation, including Walter McCarthy's own analyses of the past. There was nothing comforting about his reminders:

> Without the capacity for improved heat removal, melting of the fuel element alloy begins 27 seconds after the onset of an accident.

Or:

> Fast reactors gain reactivity if the fuel is rearranged in a denser condition.

Here was a delicate situation. The alarms had gone off. Radiation was leaking. Some core temperatures were inexplicably high. Direct inspection of the core of the reactor vessel was impossible. Even if the containment building were not sealed off, there was no way to see if the fuel rods were melting, how much damage to the core had been sustained, what direction the accident was taking, or the shape of any melted fuel.

There was plenty of past theory to go on, both from McCarthy's previous analyses and others. J. R. Dietrich had written in the *Technology of Nuclear Reactor Safety*, a nuclear engineer's Bible:

> Any accident which can cause a compaction of the fuel may produce a very serious increase in reactivity.

But was there compaction of the fuel? Would there be? Decisions would have to be made quickly, and they would have to be made carefully. A wrong decision might be worse than none at all.

The maintenance engineer, Ken Johnson, was at his desk in

his office, not far from the control room when the alarms went off. He ran down the short corridor to the relay room, where there was a panel of gauges monitoring the radiation levels. They were reading high, especially in the containment building. His first thought was whether there was anyone in it. No one could enter without clearance. The only entrance and exit to the containment shell was through an enormous double door that formed an air lock. Anyone entering would have to step into a chamber and wait for the outer door to close. Then the inner door—as thick and enormous as that of a huge bank safe—would open. The process was timed for thirty seconds. Johnson picked up a phone and called Bob Carter, his maintenance foreman. He asked for an immediate count of the crew. They were all present and accounted for.

Johnson quickly scanned the possibilities of what had caused the radiation alarm. One thought was that one of the seals that kept the argon gas from leaking into the containment building had failed. The argon gas was critical. It sat invisibly inside the reactor vessel and kept the oxygen from hitting the sodium, which would flash into a fire or an explosion if these two elements combined.

At this point, the reactivity rate was unclear, and the situation was confused. Johnson's thought about the argon gas was that it would have some radioactivity in it under normal conditions. Of course, if the fuel had melted, it would be highly radioactive, as the fission products would have burst out of the spalled fuel cladding and saturated everything in the reactor core with their poisons.

At the control room console, the operator had begun to pull down the power as soon as the radiation alarm sounded, dropping the rods slowly to see if the reactor could be brought under control. No one knew yet what had happened, or why it had happened. There were almost endless possibilities, and any decision had to rest on a careful assessment of all the instruments.

On the panels in the control room, there were over two hundred dials, gauges, and warning lights alone—not counting those on the control console.

A natural impulse, of course, would be to scram the reactor immediately. But there were problems with this. Thermal shock, due to sudden changes in the sodium temperature, had to be guarded against, in both the blazing hot core and the channels that carried the coolant.

This sort of problem left the operating crew between the devil and a runaway meltdown. Yet how could any engineer or reactor operator be cool enough to handle the complexities in a crisis situation? Even if a technician memorized every factor, every golden rule laid down in the industry's Bible, how could they be correlated in the seconds—or minutes, if they were lucky—that were allowed in a nuclear accident crisis?

Mike Wilber was still trying to put his finger on what was happening inside the reactor core; he was checking and rechecking the instruments. So far, at least, the radiation was not threatening the control room, and it was within reasonable limits where it was coming out of the tall stack. It had not yet reached intolerable limits outside the containment shell. Already, a team of health physicists under John Feldes was circling the outside of the containment building with Geiger counters. The readout on the fission product monitor—which later proved not at all reliable—showed moderate radiation around most of the area, but there were high radiation levels near the number 1 pump, and the area was roped off.

In an emergency situation such as this, time is crucial; confusions and complications create frustrating delays. One complication was that a Fermi instrument engineer had been working on the fission product monitor, checking the calibration on the panel. When he saw the steep climb in radiation at the time of the alarm, he thought immediately that he had merely triggered a false alarm while working with the instruments. But the temperature readings on the subassemblies and other indicators showed that something was happening in the core that was very real. And so at 3:20 P.M., eleven minutes after the radiation alarm had gone off, the decision was made to manually scram the reactor. The question: Was this too soon or too late?

All the rods went down into the core normally, except one. It stopped six inches from the full "down" position. This was no time to take a chance. A second manual scram signal was activated. The reluctant rod finally closed down fully.

Ken Johnson made his way to the control room. The red light in the corridor which had read REACTOR ON, was no longer on, so he knew now that the reactor had been scrammed. The control room was quiet; operators for the new shift were coming in; several staff members were checking instruments and charts, trying to find out what the trouble was. Johnson knew immediately that it was something serious, something a lot more than a faulty seal. He found Mike Wilber very concerned about the situation, especially the high temperature readings in the core. All the signs seemed to be pointing to a fuel melting situation, and there wasn't a nuclear engineer in the business who didn't know what that could mean.

Walter McCarthy was in a conference in downtown Detroit when it happened. He got a call from Bill Olson, the plant supervisor, who told him that there definitely was evidence of fuel damage, that the reactor had been scrammed, and that the containment building had been isolated with high radiation levels. McCarthy called his wife to say he wouldn't be home for dinner. Then he tried to reach Walker Cisler, who was in New York at the time. He couldn't reach Cisler, so he took off immediately for Lagoona Beach.

When he arrived at the Fermi control room, there was still confusion as to what had happened. The critical question remained: Was there fuel melting or not? With direct observation impossible, the problem would boil down to instruments, deduction, and a prayer. The only hope for future inspection was to drain the thousands of gallons of the thick, opaque sodium out of the reactor, and then, with infinite care, to try to probe the bowels of the core to see what had happened. This was, of course, impossible at the moment.

McCarthy didn't need to be reminded of the words of J. R. Dietrich in the nuclear engineer's Bible:

> In all but the smallest and most compact fast reactors, the agglomeration of even a fraction of the total fuel into a compact mass will usually result in a highly super-critical assembly. . . .

Some kind of fuel melting was suspected by Mike Wilber, and if his theory was correct, the melting could be in more than one fuel subassembly. The question here was: How much fuel had melted, what was the condition of the core, and what were the chances of a secondary accident?

Again, Dietrich had given a very clear and terrifying picture of this:

> In a fast reactor, the dynamic portion of a reactor accident cannot be considered to end with the general melting or thermal failure of fuel elements. On the contrary, it is conceivable that the serious portion of the accident may only begin at that point.

It didn't take long to deduce that there was definitely fuel melting, and that it wasn't confined to a single subassembly. If there was melting in several subassemblies, it would create a situation that would require extreme caution.

Almost immediately after he arrived at the plant, McCarthy called a meeting. Every available key man of the Fermi team was there—Olson, Wilber, Jens, Amorosi, Johnson, and others—some of whom had nursed the plant from its infancy, for over a decade. Alexanderson was to arrive later.

The prime questions were: Is the reactor secure? Would it stay secure? What could be done to explore the accident that wouldn't trigger a secondary accident more terrible than the first? The urgent, burning priority was to make sure that no hazardous condition existed in the core. The potential hazard was of course enormous, and the lack of experience in handling fast breeder accidents made the situation fraught with danger. Further, no provision had been made in the design for investigating and recovering damaged fuel elements.

To say that the Fermi team was sitting on top of a powder keg would be a major understatement. The threat of a secondary accident was, as McCarthy was to say later, "a terrifying thought. . . ."

However terrifying the situation, it was staring the Fermi crew in the face. The keynote was *uncertainty.* There were few road maps to go by. No one at the hastily called meeting knew exactly what had happened within the reactor core. No one knew what would happen if they tried to look inside it—or how to look inside it. The most probable cause of fuel melting was the blocking of the sodium coolant.

McCarthy took command by saying: *"We will go at this very, very, slowly."* Before any kind of exploration of the condition of the reactor, a procedure would have to be written. It would have to be checked and double checked before any attempt to put it into action would be permitted. Again, there could be no margin for error.

Outside of the tense atmosphere in the Fermi plant conference room, there was no outward sign of trouble at Lagoona Beach. Speculation about a peacetime nuclear accident had been kept in such a low profile by the AEC that hardly anyone would be likely to think about it. A coal mine disaster, a chlorine explosion, an ammunition ship blowing up—all were tragic sorts of things that could happen. But none of them threatened to contaminate a whole state or to kill in such potentially massive quantities. None would threaten the soil, the vegetation, the water tables, the air for thousands and thousands of years.

There were a couple of public laws in Michigan, dating as far back as 1953, which provided for the attempt to cope with such a possible catastrophe. The department of public health was named the official radiation control agency. But how could a cloud of radiation that could fan out to cover an area the size of a state be controlled, even by the most expert public health officer? The department of state police was designated as the coordinator of civil defense activities if and when the governor proclaimed an emergency. But how could a handful of state police handle a gigantic exodus from the city of Detroit or even from Monroe County?

The scene was almost unimaginable. Trucks, cars, buses stalled in massive traffic jams along the superhighways. Long streams of people carrying blankets, pots, pans, children moving out of Detroit toward Ann Arbor, Lansing, Grand Rapids, Ontario—themselves places of dubious safety under the silent plume of radiation. And yet, in the AEC meetings at Brookhaven the only answer that had come up in the discussions was evacuation.

The state of Michigan plan reads with simple eloquence:

> In the event that an incident occurs which releases radioactive materials in concentrations that may be a public health hazard, this plan will be implemented. Implementation will commence by proclamation of an emergency by the Governor or by the order of the Director of the Department of Public Health. . . .

This department will:

> perform monitoring, evaluate data, and establish emergency response actions.

But what would these actions be?

The Michigan radiation emergency plan had many provisions. One of them was that the state police would notify the bordering states and provinces of the approaching danger. But what in turn could these states or provinces do, aside from the vagaries of "establishing emergency response actions?" What is the "response action" for a cone of radiation that will settle an invisible mantle of contamination only God knows where?

The plan included neat and tidy classifications of exposure conditions. There were three main classes: "Whole body exposure, including eyes, gonads, and blood forming organs," exposure to "the thyroid of a one-year-old child," and "liquid discharges." The only real answer was to vacate the area.

The plan itself seemed an exercise in futility. But so was the meeting in the conference room of the Fermi plant. The few confirmed facts that could be accurately determined at the time were that, because of the combined readings of several meters, there had not only been fuel melting, but there had been "fuel redistribution," meaning that the fuel had shifted as well as melted. This would automatically leave the way open for further and more serious accidents to happen. And there was still the question as to whether the reactor had been scrammed soon enough.

McCarthy, still trying to reach Cisler on the phone, directed the meeting toward getting at the possible cause of the accident. He was afraid now that the two hot subassemblies were not the only ones that had melted. But he could not be sure because only one in four had temperature gauges throughout the core. It would be sheer luck if M-140 and M-098 were the only ones involved.

Many explanations of what might have happened were brought up at the meeting: broken fuel pins, strainers, foreign material on the pins, fuel swelling, and other possibilities that might have blocked the coolant from coming through the subassemblies. Somehow, somewhere, the melted subassemblies must have been starved from their protective sodium coolant, either by some foreign matter that blocked the nozzle, or by the flow behavior of the sodium itself.

McCarthy laid down two programs. One was to work out a detailed analysis and experimental program to find out just what the chances of a secondary accident could be. The other was to try to find out what the cause was, and to try to get the reactor back into service.

But the first problem would be the one hanging over not only the heads of the crew, but the entire state of Michigan as well.

FOURTEEN

In addition to facing "hair-raising decisions" and terrifying thoughts, the men who sat in the control and conference rooms at Lagoona Beach also faced a deep sense of concern and frustration. Practically all the rules in the book warned of a secondary potential that could be far worse than the original melting. And the slightest disturbance of a partially melted core could easily set off that more powerful secondary accident.

Worse, the primary accident itself had gone beyond the confident predictions for a "maximum credible accident" of both McCarthy and Hans Bethe. McCarthy had stated flatly that only one subassembly could melt in the Fermi reactor. The instruments already showed that two were affected, and there were probably more. He had also stated that the reactor would shut down automatically in such a situation. It had not; it had required a manual shutdown. Hans Bethe had testified that a core meltdown accident was "incredible and impossible." Both were experts, and both were wrong.

Now, with the reactor shut down, and no one knowing what

possible shape the core and the melted uranium were in, what could be believed about the other predictions?

Cisler, finally contacted, remained calm in the face of the tragedy. McCarthy recapitulated what had happened, trying to deduce why the accident had occurred, but without having much real evidence to work on.

With the belated scramming of the reactor, the radiation leakage had begun to drop off. It was some comfort, but the concern was what might now happen in the core. The only way to get at the core was through the fuel-loading contraption, the awkward and clumsy Lazy Susan mechanism that provided no vision of what was going on, and that could easily jar a partly melted core into a secondary accident. It was like trying to look inside a gasoline tank with a lighted match. How could they explore a reactor drenched in radioactive poisons without the risk of wiping out Detroit and a big chunk of Michigan with it? Ironically, hardly anyone in Detroit, or the state of Michigan, had any idea of the potential danger they were in.

As the afternoon of October 5 wore on, Sheriff Bud Harrington sat in his miniature office in the Monroe Town Hall, but no further phone call about the incident came in. Captain Buchanan of the Michigan State Police heard no more about the alert that day either.

Frank Kuron, the barrel-chested ironworker, was in his living room in Stony Point, two miles away from the reactor, his feet propped up on a footrest, watching the Baltimore Orioles take the Los Angeles Dodgers in the first game of the 1966 World Series by a score of three to two. Pitcher Drabowsky was having a good day, striking out eleven batters in the process. It was a good day for the Polish, Kuron was thinking.

Other Michigan citizens were equally unconcerned. No news of the accident had been broadcast that afternoon or evening, but McCarthy made it a point to call the local Monroe paper to give out a brief but very ambiguous statement to appear the next morning. The communique was couched in terms reminiscent of a wartime battle report. It mentioned nothing about any fuel melting, merely stating that the "radioactivity level of the argon gas" was "observed to rise substantially." It went on to mention that this resulted in the automatic sealing of the building ventilation system. The release was a masterpiece of understatement, but it would get McCarthy off the hook if rumors began spreading around town that something was awry at Lagoona Beach. The irony of the situation was that even if any citizen in the area had known that a meltdown was in the works at the Fermi reactor—which no one did—there would be little alarm. The situation was too new, too obscure, too unfathomable for anyone untrained in the overwhelming technicalities of nuclear engineering. It was a case of ignorance not only being bliss, but comfortably reassuring.

In one way, the understatement might have been good in the light of what was to happen in Sweden several years later. One November day in 1973 an enterprising producer for Swedish radio decided to produce a fictional drama about an actual atomic power plant in southern Sweden. He wanted to point out to the public what a horrible catastrophe it would be if the reactor went into a major accident. Just as Orson Welles had unwittingly done in his notorious radio dramatization of *The War of the Worlds* back in the 1930's, the Swedish producer shaped his radio drama so that it sounded so vividly real to both Swedish and Copenhagen listeners that most of the population went into shock. Panic hit suddenly in southern Sweden—where the fictional meltdown was described as taking place—and much of Denmark across the way.

Every hospital in hundreds of square miles went into a full alarm situation. The population rushed to fallout shelters. Fire stations called in emergency crews. The phone system in both Denmark and Sweden broke down within half an hour. In Copenhagen, doctors rushed into the main hospitals to stand by for supposed radiation victims. People with trucks began loading them up with furniture and possessions. Householders rushed to seal off windows and doors. Within ten minutes, the roads were jammed with refugees, traffic was in a hopeless tangle, and total panic had set in. Even reassurances and disclaimers that followed the broadcast failed to calm things down.

It was a tragic scene, all the result of a new Orson Welles situation—but it dramatized powerfully what could happen if a real reactor accident were announced in a heavily populated area.

But the accident at the Fermi site was very real. Even though the control rods were shoved down into the core, there was no assurance that a secondary accident could not take place. This was the problem the Fermi engineers faced—and what Detroit and its surroundings would face if the worst did happen as a result of probing into the causes.

Under these conditions, there is little wonder that the Fermi engineers talked of "hair-raising decisions," and "terrifying thoughts." In effect, they were sitting on top of a volcano, which, if left alone, might be all right; but, if they tried to take a peek inside, it might erupt. Yet they were forced to take some action. They could not walk away from the accident, even if they wanted to. Aside from the hundred-plus million investment at stake, it would be impossible to leave a hot reactor sitting there, loaded with deadly fission products, soaked in radioactive sodium, and choked in an unknown configuration of melted uranium. The reactor could not be sealed in a tomb of cement and forgotten about. Sooner or later, its poisons would eat down through the base, no matter how thick, and contaminate the water table and soil below it. A cement tombstone, even if poured lavishly over the core, would eventually sweat out the buried poisons and continue the contamination for thousands of years. Not only that, but if the concrete became wet when it was irradiated, there could be what is called a "radiation dissociation" of the water to hydrogen and oxygen which would actually create an explosive mixture, turning the entombed reactor into a bomb. It was too late to go back to the drawing boards.

The lights burned all through the night at Lagoona Beach as McCarthy, Cisler, and the others pondered over the situation. They kept a close eye on the weather situation. That afternoon the wind had been blowing from the west, and if the unthinkable had happened it would have swept the radioactive cloud out over Lake Erie, losing some of its radioactive punch by the time it reached Pennsylvania to the east. However, by nightfall, there was considerable nocturnal inversion—the worst kind of condition for a deadly plume of radiation. And by the next day, the wind had shifted to 220°, a course that would take any radioactive fallout smack into the lap of Windsor and much of Detroit. Under these conditions, especially, a secondary accident with breach of containment could not be allowed to happen.

It is a strange thing that in a situation of harrowing, unthinkable tragedy and drama an atmosphere of silence and calmness can take over and dominate the scene. With the realization of the potential for death and destruction, no one could fail to be nervous. But practically no one showed it. Emotions were deep-frozen in the coils of the engineering and scientific minds. It was a deceptive calm that permeated the control room as the second day of the meltdown began.

For several days, the only thing the Fermi crew could do was gather ideas, and write up procedures as to how to carefully carry them out. But the time bomb was still ticking, quietly and relentlessly. During those days, the weather grew less and less cooperative, with the wind shifting so that any escape of radiation would cover the maximum population of Detroit and its sprawling suburbs. The day of the accident marked the beginning of a warm spell, so any escaping radiation would tend to hang lazily under the nocturnal inversion conditions that existed through each night—the warm air trapped by the cool lid that boxed it in. It was, in fact, perfect Indian Summer weather—the worst possible kind for any ground release—a time when the haze from burning leaves hangs heavy and stagnant in the air, creating a smoky veil that lingers pungently for days.

The most frightening thing for the scientists and engineers was not knowing the cause of the trouble. There were guesses at the meetings as to how long it would take to find out what had gone wrong inside the bowels of the reactor. Some figured it would take a year, if all went well and no secondary accident took place. Some guesses were that it would take even longer.

Finally they decided to run some cautious tests. In one tense and timid exploration, the control rods were withdrawn one at a time to test the reactivity situation, and then shoved back into the core. On reading the instruments, it was confirmed that fuel melting had definitely taken place.

But there was still no clue as to why. They carried out another check of the rate at which the liquid sodium flowed. By deduction, the test showed that no more than six complete subassemblies could have melted. In checking the flow of the sodium, a microphone was placed cautiously on a control rod extension to see if any clue might turn up from the sounds inside the reactor. A clapping noise was immediately detected. It slowed down when the flow of sodium slowed. But this offered little enlightenment.

The reactor had now been closed for a month, and the endless meetings and cautious tests still had produced no clue either to the cause of the problem or its solution. Tired, exhausted, frustrated, and concerned, McCarthy looked to the outside for suggestions. The situation demanded the best brains in the nuclear field, and he set about getting them. They arrived in Detroit from all over the world: from France, from England, from Scotland, from all over the United States, specialists and experts who would try to hold a conference and diagnose a patient that was unable to be viewed or even poked at.

The international "medical" consultation lasted for hours. Every aspect of the patient's condition was surveyed. It was agreed that, in view of the time that had elapsed, some bolder action might be taken to explore the secrets held inside the core.

They even considered trying to remove the damaged subassemblies. Knowing the dangers of a fuel-loading or fuel-removing process, even under normal conditions, it was not a step to be taken lightly. The second accident at Chalk River, Canada, where only a single piece of unenriched uranium had almost caused a major disaster, had dramatized that. The Fermi fuel was highly enriched, tightly packed, and much more threatening than the simple, chunky piece of Uranium-238 that had so thoroughly ravaged the NRU reactor at Chalk River.

A decision had to be made. The engineers could no longer wait and watch. Inside the bowels of the reactor vessel was a maze of unknown geometry. Exploring it could be disastrous if the wrong steps were taken. A slight jar or bump under the wrong conditions could catapult the reactor into a helpless runaway. But somehow that wrecked fuel had to be hauled out, and the unknown faced.

Finally, a consensus was reached. A surgical incision had to be made, and the scalpels prepared for an operation to remove the diseased fuel.

The situation changed from one of trepidation to one of crisis.

There was practically no experience to go on, and any attempt to remove melted fuel from the reactor was fraught with catastrophic potentialities. To get at the damaged subassemblies meant raising the giant hold-down device that sat on the top of the core like an enormous spider. If any of the fuel subassemblies were sticking to its huge fingers when it was lifted, there could be hell to pay.

In an atmosphere of controlled tension, the Fermi engineers began the job. Using every measuring instrument in the core that could apply, the hold-down device was raised very slowly. It worked. Apparently nothing was stuck to its claws. This was a major source of relief. Then, very slowly and cautiously, a mechanical arm was swept over the top of the core to check on whether any of the subassemblies were poking up above the top level of the core. Again, there was success. The chances of a secondary accident appeared less with each step.

Then the big lobster claw that would sweep over the core to lift out the subassemblies was brought into action. A special weight gauge was installed on it. All this was being done blindly of course. There was no way the reactor shield tank could be opened to look inside. It was blistering with radioactivity, and filled with the argon gas that kept the sodium away from the air.

The idea was that any subassembly that had melted would weigh less than the normal ones. In this way, the damaged fuel could be located, and with luck, removed and examined far away from the reactor in what is known as a hot cell. Here, everything is handled by remote control, behind heavy shielding and lead glass windows up to four feet thick. For this process, the damaged fuel would have to be shipped to Columbus, Ohio, in special casks. This alone was a formidable job, facing the constant bugaboo of a transportation accident that in itself could be deadly.

The detective work that began a month after the accident continued week after week, month after month, at a snail's pace. It was essential to learn exactly how much fuel had melted to eliminate the possibility of a secondary critical accident in case some of the fuel shifted during the exploration process. The investigators finally learned that not two, but four subassemblies had been damaged, with two of them stuck together.

It took from October, 1966 to January, 1967 to determine this, and from January to May, 1967 to remove the damaged subassemblies. Removing them was a precarious and overwhelmingly difficult five-month-long job. Special optical devices and cameras were devised. Part of the thick, opaque sodium syrup had to be drained from the reactor, although there was no provision for this in the reactor's design. A shielded viewing window had to be inserted in the plug at the top of the vessel. A borescope placed on the end of a flexible tube was pushed down into the reactor.

By August, 1967, more of the sodium was drained out to expose the meltdown pan at the very bottom of the reactor vessel. So far, even the warped and twisted subassemblies gave no clue as to the cause of the accident—an accident, they said, that could never happen. By September, nearly a year after the meltdown, they were able to lower a periscope through a stainless steel pipe that was shoved down through a hole in the plug that circled the top of the reactor vessel. A quartz light was rigged to slide down it. The device finally reached the meltdown pan, forty feet down, at the very bottom of the reactor. There, the inverted ice-cream cone, known as the conical flow guide, sat as the alleged guardian to spread out any melted uranium that had spilled down onto the meltdown pan.

As the periscope scanned the bottom of the vessel, it became apparent that there was no melted uranium there. But there was something else. Manipulating the forty-foot-long periscope and light, the engineers saw what looked for all the world like a crushed beer can, lying innocently on the meltdown pan. Here, at last, could be the cause of blockage of the coolant nozzles of the subassemblies; a flattened piece of metal that could easily starve off the sodium and allow the uranium to melt, the cladding to rupture, the subassemblies to warp and twist, and the fission products to burst out.

But how did the beer can get there? Had some worker

carelessly dropped it from his lunch pail and unwittingly nullified all the carefully planned safety devices that would protect against a meltdown? And *was* it a beer can? And if not, what was it? The detective story wasn't over yet.

As the Fermi engineers worked and sweated to get at the mystery, the critics began firing salvos at Cisler, McCarthy, and the rest of the crew. Sheldon Novick, a concerned environmentalist of Washington University in St. Louis, and editor of the magazine *Science and Citizen,* hit hard at the Fermi project in his magazine when he wrote that the accident far exceeded the worst envisioned, and could have meant disaster for citizens in the Detroit area. "The huge quantities of radioactivity involved and the proximity of Detroit made the prospect terrifying indeed," he wrote. Then he continued: "It should be emphasized that the maximum credible accident was assumed to occur at a power level 15 times that at which the actual accident occurred. In other words, the actual accident was not only 'incredible,' it might have been far worse." Novick concluded that the only answer was to shut down the Fermi plant forever.

McCarthy rose to his own defense and claimed that there was no danger at any time. George Weil called the Fermi reactor "a costly project which might have ultimately led to an explosion and release of radioactive elements" with a "second accident potentially catastrophic to surrounding populated areas." Leo Goodman, another critic, accused McCarthy of "taking reckless chances with lives in the Detroit area," and added: "What happened was precisely what we have been predicting since 1956. If they continue to operate the plant, they are likely to have another meltdown and nuclear runaway—an uncontrolled reaction." In fact, Goodman strongly opposed his daughter studying law at the nearby University of Michigan because Ann Arbor was too close to the Fermi site.

Whatever the critics were saying, McCarthy and Cisler were determined to uncover the mystery of the so-called beer can, and get the plant back into operation. Cisler, however, was subdued and seemed to have lost his fire. He said: "The Fermi plant will be more valuable as a research facility than as a source of electric power." Critics pointed out that, up until the time of the accident, more than $120 million had gone into the project, and that the Fermi plant had been able to produce only fifty-two hours of electricity over a ten-year period. It had been able to produce no plutonium fuel whatever.

Now the reactor had lain fallow for a year since the accident, with no hope of operating until the beer can—or whatever it was—was removed from the bottom plenum and months of checking and repairs were done if indeed that was at all possible. The attempt to retrieve the metal object and several other scraps of metal that showed up in the remote-controlled photographs was described by McCarthy as: "Like taking out an appendix through the nostrils." It was a laborious, painful, frustrating job, requiring manipulations and dexterity that the reactor was never designed for.

The first problem was to identify the crushed metal more clearly. There were fifteen optical relay lenses in the pipe that reached to the bottom of the reactor. The first pictures failed to give enough clarity. Somehow, they had to find a device that would shove the object nearer the camera, and flip it over so they could get a better view from all sides. To do this, it was necessary to cut a hole in a coolant pipe halfway up the side of the vessel, with two right-angle turns in it. A sealed glove box was put over the hole to keep in the radiation, while Ken Johnson got to work to design a flexible, bicycle-chain type of tool to slide down the second pipe, in the attempt to bring the object nearer the periscope. It was a delicate fishing expedition. Terminating the flexible, bike-chain tool with a short piece of wire with a small hook at the end seemed to work best.

"We could then pull the piece of metal closer to the periscope, turn it over, and photograph it from all angles," Ken

Johnson said. But even these pictures, viewing the object from many angles, failed to give a clue as to what the object was. The beer-can theory seemed less and less probable, however. Nuclear scientists everywhere in the country were asked to look at the pictures, and they agreed on one thing: The piece of metal bore no resemblance to anything used in the construction of the reactor.

This failed to faze McCarthy. "We're no longer concerned as to how the metal piece got into the reactor core," he said at the time. "The problem now is to get the piece of metal out of the core, and get the plant back into operation as soon as possible. We're pretty well satisfied that it may have entered the base of the core through the sodium intake pipe, and then was carried by the sodium against the base of the subassembly tubes, shutting off the flow of the coolant and causing the subassemblies to overheat."

But with more special tools to be designed, and with safety still a major concern every step of the precarious way, McCarthy knew that there would be no chance of getting the reactor back into action for many months. His most optimistic guess was that the middle of 1968, nearly two years after the accident, would be the earliest time that Fermi could run again. Meanwhile, the identity of the crushed-metal culprit and how it got into the reactor remained a mystery.

News of the Fermi meltdown was kept quiet, but it continued to attract the attention of critics. The scientists among them knew that, no matter what the Fermi engineers would admit to in the way of terror or concern, it could be only a fraction of what their inner feelings were. If they weren't scared to death when the accident happened, they would not be human. And, as 1968 began, the critics gathered more artillery against the Fermi breeder—to say nothing of the whole concept of any fission atomic plant.

George Weil, one of the most competent and qualified of the critics, amplified his position this way: "Under current plans for the accelerated growth of nuclear fission to meet our energy requirements, we are committing ourselves to the nightmarish possibility of a radioactive-poisoned planet. Today's nuclear power plant projects are a dead end street. There are too many, too large, too soon, too inefficient; in short, they offer too little in exchange for too many risks. The commitment of billions of dollars to the development of breeders . . . would almost certainly be an irreversible decision, foreclosing any serious consideration and adequate federal funding of alternative energy sources. . . . With determined efforts to harness fusion, solar, and other energy systems, we may well escape the threat of ecological radioactive disaster."

Meanwhile, McCarthy's hope of fishing out the mysterious piece of metal, and getting the reactor back into action by the middle of 1968 was fading every day, as the retrieval process demanded more and more engineering ingenuity.

With only two ways to reach the bottom of the reactor—the long straight tube down from the top, and the double-bent sodium coolant pipe, teams had to work at each of these outlet points, coordinating their efforts together. The long, straight tube served as the periscope viewing station from forty feet above the base. The bent sodium pipe served as the channel for the special tools. They were designed to snip samples of the crushed metal, pull them up through the tube for examination and ultimately to identify them. The tools, created at enormous expense, were dubbed such names as the "hawk-bill cutter," and the "organ pipe tool," and were of unbelievable complexity. The shifts worked twelve hours a day, fishing down through the two pipes and manipulating the tools and light by remote control. Ken

Johnson worked from the top of the reactor, forty feet up, while Phil Harrigan, his assistant, worked the snake-like tool from the side of the reactor, thirty-five feet away from the base. An intercom system kept them in touch. All the work was done in specially-built locked-air chambers to avoid radiation.

Johnson worked the quartz light like a hockey stick, trying to maneuver the elusive hunk of metal into the jaws of Harrigan's retrieval tool, with Harrigan working blind and following Johnson's instructions. The lenses in the periscope would cloud up, and it would take a day to clean them.

Finally, at 6:10 P.M. on a Friday night, almost a year and a half after the meltdown, the crushed metal was firmly gripped by the tool Harrigan was operating. It was drawn slowly up the coolant pipe. It took an hour and a half to lift it up. With the temperature inside the reactor at 350°, the metal was given time to cool as it was drawn up to the surface.

It was examined carefully, and finally identified. Ironically, it was one of the five triangular pieces of zirconium that had been installed as an added safety measure to protect the upside-down, cone-shaped flow guide. Somehow, one of the five pieces had worked loose and clogged the coolant nozzles. It had been installed back in 1959 and forgotten. It wasn't even on the blueprints.

FIFTEEN

The "terrifying thoughts" about the accident at the Fermi plant did nothing to squash the enthusiasm of the supporters of fission power who countered rational questions with flimsy answers. Was there no possible way to guarantee an emergency cooling system for the light-water reactors? Build them anyway, and hope for the best. Was the breeder reactor erratic, dangerous, unproven, and untamed? Forget Fermi, and plan another one, a bigger one, in Tennessee. Was there absolutely no solution whatever for the safe and eternal burial of radioactive wastes? Worry about that later. Was there no safe way to transport spent fuel by truck or train or air? Ship it anyway, and take the chance.

This was the situation in April, 1968, as McCarthy and his crew were trying to figure out how they could clean up the Fermi reactor and get it working again. But the accident had already lost them eighteen months of valuable time and the outlook seemed dim. Estimates for the clean-up job ran anywhere from another year to a year and a half—if not longer.

McCarthy displayed some signs of humility when he said

late in the autumn of 1968: "Fermi is a long way from being a financial success. It can never be economically competitive. It is an experimental reactor, built so that we could learn the engineering and economics involved. We have learned enough and trained enough men to justify the project." But aside from admitting that, with the Fermi project, "perhaps we were a little ahead of our time," he boldly told reporter Chester Bulgier of the *Detroit News* two years after the Fermi accident: "The breeder reactor is the world's hope for increasing energy to meet the world's needs, because it can make more fissionable fuel than it consumes." The Edison Electric Institute, a trade organization consisting of all the important industrial and utility nuclear energy groups, backed up this position. In a report issued at the time, the industrial group urged the development by 1985 of a fast breeder more than twice Fermi's size.

Milton Shaw of AEC's division of reactor development also concurred. "I'm more convinced than ever that this is the way to go," Shaw said. "I'm not discouraged at all by the problems Fermi has encountered because they are problems which are amenable to engineering solutions. The fast breeder sodium plant will be a tremendous boon to mankind."

Shaw obviously meant what he said. He continued to push for bigger budgets each year, nearly half the total AEC budget going into breeder development. Billions of dollars were being ladled into fission reactors of all types—with the breeder program jumping to nearly half a billion a year by the mid-1970's.

With such lavish funds becoming available, it did not take long for Cisler to get his mind set on another breeder reactor at Lagoona Beach, one that would be several times larger than the ill-fated Fermi plant. And despite the contention that no experiments should be done in a populated area, he was promoting a major role for the old Fermi reactor—the irradiation and testing of new types of nuclear fuel.

McCarthy agreed with him on the experimental fuel

question. "We are really using antique fuels now," he said. "A plutonium core is being developed for Fermi, and even more potential is shown by uranium and plutonium oxides."

McCarthy's engineers moved ahead, straining to give the Fermi reactor artificial respiration, painfully hauling out hot fuel subassemblies one by one, checking them over, throwing some away, keeping others. It was not until the last weeks of 1968, more than two years after the accident, that the remaining pieces of the zirconium "safety" plates that had set the meltdown on its way were fished out of the reactor. Then it took until February, 1970, more than three years after the accident, to get AEC permission to reload the reactor with new fuel.

By May of 1970, the Fermi No. 1 plant was nearly ready to resume operation. With the AEC inspectors prowling everywhere, extra care and caution were being tendered to the process. Nerves were frayed. Workers in the 10-by-10-foot room holding the sodium transfer tank were meticulously preparing that phase of the operation when suddenly two hundred pounds of radioactive sodium burst out of the pipes. Other pipes broke loose and doused the sodium with water. The sodium flashed immediately, then exploded. Argon gas was rushed into the building. The fire and explosion were contained, but no one could enter the building for two days, and then only with extreme caution.

Again luck held. With determination and obstinancy, the Fermi engineers pushed on. In the middle of July, 1970, the plant was fired up again, and a whole new series of tests were begun. Slowly, they began pushing toward the re-licensed 200,000-kilowatt heat level. By October, they had reached it.

But at the same time, the Michigan Public Service Commission was taking a hard look at the operation. Not only were there safety problems, but Detroit Edison was pouring more money into the shaky and wavering Fermi project at the expense of both its utility rate payers and its investors. Moreover, the other corporations that had so enthusiastically joined the Power Reactor Development Corporation were now backing out. The risks, both financial and safety, were too high. The only one who seemed eager to try to hoist Fermi out of its financial morass was a Japanese organization that was willing to pay cash just to learn the business.

To keep the plant on its feet, Detroit Edison was paying the AEC $65,000 a month for the use of the uranium fuel. The amount of electricity produced since Fermi began operations was practically nothing. Even running at its capacity, the power produced would cost up to fifteen times as much as coal. Its only *raison d'etre*, therefore, was that of a demonstration plant. Even so, it was doing very little demonstrating. The total costs for the project had now edged up to $132 million, and in January, 1971, the AEC license was due to expire.

By the license expiration date, no one else would put up any more money. Grudgingly, the AEC extended the license until June of 1971. It dropped the fuel charges, which were running up close to $750,000 a year.

The situation was complicated by Detroit Edison's plans to build a light-water fission reactor—to be called Enrico Fermi Plant Unit No. 2—next to the dying Fermi No. 1 breeder site, and later, a third one was to join it. The amount of fission products that these three reactors together could generate would be almost beyond comprehension.

This fact, combined with the total lack of a proven emergency cooling system for light-water reactors, was ominous not only for Detroit, but for almost any other populated area in the country where the same thing was happening. According to the plans created in 1971, there would be more than fifty of these giant light-water reactors scattered around the country by 1974. From then on, the pace would quicken. Almost one thousand atomic power plants were planned for the end of the century. And while the problems with the light-water reactors were different from those of the sodium-cooled breeder, they were just as serious. If a major pipe broke, and if the cooling water was lost, there could be a catastrophe. There wasn't a nuclear engineer in the business who would deny it.

A senior AEC engineer at the Oak Ridge Laboratory told Robert Gilette, who was writing a series of articles for *Science*, the prestigious publication of the American Association for the Advancement of Science: "What bothers me most is that after 20 years we are still making purely subjective judgments about what is important and what is not in reactor safety. Purely by decree some things, like the rupture of a reactor pressure vessel, are ruled impossible. To decide these things without some objective measure of probabilities is, to me, almost criminal."

He wasn't alone in his thinking. Another Oak Ridge scientist, Philip Rittenhouse, compiled a list of nearly thirty of his professional associates "who consider the present safety standards seriously deficient in a large number of fundamental technical assumptions."

Frank Kuron agreed. He was back working at Lagoona Beach—this time on the red iron of the new reactor, Fermi No. 2. But in contrast to the workmanship he had admired in the Fermi No. 1 reactor, he wasn't pleased with the new job. No one quite seemed to know what he was doing, Kuron felt. It wasn't like the old days, a decade and more before, when the welding was clean and tight.

But worse than that, there was trouble with the excavation for the new reactor. Unlike the neat, dry hole for Fermi No. 1, the new one was flooding with water. Wells for all the homes in Stony Point were running dry. Blasting of rocks was splitting cement doorsteps and plaster in the area. There were problems with the steel reinforcement rods for the new containment building.

"They had to be right on the money," Kuron said, "but they weren't. In fact, we had to tear out eight rods that were an inch or so short. We had to take them out before we poured that floor. Okay, what happened in the end was that this floor started to crack up. There was so much water around, they had two de-watering pumps going twenty-four hours a day. The damn building could have been floating away. They tried high-pressure grouting. It didn't work. So we got a floor full of cracks."

It was at this time that Kuron ran into Tom Morgan, a lean, laconic auto worker, from West Virginia. Morgan was a trustee on the Frenchtown Township Board, and a shrewd, intuitively intelligent maverick. Like Kuron, he had little formal education, but his vocabulary and insights were impressive. He was extremely interested in what Kuron had to say about the workmanship on the new Fermi No. 2 reactor, because he had been boning up on the entire atomic power plant picture in line with his responsibilities to the local citizens. He had managed to read through, thumb, and underline a yard-high stack of hearings of the Joint Committee on Atomic Energy, and could discuss highly technical problems with the best of them.

What disturbed Morgan most, as a town trustee, was the flood of complaints he was getting from the local citizenry about the continuous dynamiting at the Lagoona Beach site, and the damage to both homes and wells. But more than that, he was informed enough to realize that if the blasting was shattering cement in the homes, it could have serious repercussions at the Fermi No. 1 plant. At the time of the October, 1966, meltdown, both men had been blissfully unaware of the potential for disaster sitting on their doorstep. Now they got together to make it their business to find out just what was going on at Lagoona Beach. They called themselves The Polack and The Hillbilly, and it wasn't long before the AEC in Washington began to realize they had to be treated with respect as they began awakening the local population to the real problems underlying fission power plants.

Meanwhile, Cisler was petitioning the AEC for another extension of the license to go into a testing program for new, more exotic fuels. But by January, 1972, the Fermi plant had operated less than thirty days at its licensed capacity, for a total of 378 hours—without producing meaningful electricity or breeding any large amount of plutonium. And the AEC was beginning to mumble about Fermi No. 1 being a white elephant that would have to be replaced by another breeder in another part of the country to carry the program forward. In spite of the failures and setbacks, the AEC was determined to get a breeder going with a new design. Hopefully it would not have the temperament of Fermi No. 1.

With the fate of the Fermi No. 1 breeder still uncertain, construction for Fermi No. 2 went on, the blasting for it shattering and splitting concrete walks in the Stony Point area. By the end of April, 1972, Tom Morgan's desk was inundated with letters from an angry citizenry. Mrs. Donald Bolton of Avenue F, Stony Point, wrote: "We are being blasted off the face of the earth, but today at 12:02 PM it was terrible. We have been shaken so much, our floors are dropped down from the walls. We don't know where to turn." Mrs. Ed Whiteside of Lakeview Avenue wrote: "The plaster in my house is getting cracks in it. We have a sturdy, well-built house, but these blasts are slowly deteriorating it. If nothing is done, it wouldn't take long to get enough women to form a picket line." Other comments included: "Even our dog is scared." "The corner of our front porch started to sink." "Every time there is a blast, I find a new crack in my walls."

There wasn't much time to worry about how the blasting was affecting Fermi No. 1, however. On August 27, 1972, when the AEC issued a "Denial of Application for License Extension and Order Suspending Operations," the announcement was made that the plant would be closed forever. In the document was a sharp command to "reduce expenditures to the minimal amount practicable consistent with assuring the safety of the public and the protection of the environment, and submit a dismantling plan." There was the possibility of a hearing on the matter, but this was only a matter of form. It could not save the project.

The mood of McCarthy's staff was bitter, sad, and truculent. It was exacerbated by the AEC's decision to build a new breeder near Oak Ridge, Tennessee, then estimated at $700 million, now $1.74 billion, with much financing from former Fermi contributors. Ironically, the Oak Ridge reactor was already being referred to as "the first demonstration breeder reactor in the United States"—as if Fermi No. 1 had never existed.

Gathering cogent and coherent evidence, Tom Morgan and Frank Kuron made a frontal attack on both Detroit Edison and the AEC, appearing at hearings in Lansing, Monroe County, and Washington, D.C. What worried both Kuron and Morgan was not only the blasting, but the workmanship on the new Fermi No. 2 light-water plant. They were also worried about the radioactive guts of the old Fermi plant, which would have to be guarded forever. There were 30,000 gallons of radioactive sodium that had to be stored in metal drums, with no place to dispose of them. There was the hot core and the blanket and every piece of pipe and machinery that glowed with radioactive poisons. There was even the question as to whether the $4 million dollars set aside by the AEC was enough to bury the dead carcass of the plant. Was this, Kuron and Morgan asked, the inheritance for the future of atomic energy plants all over the country? Was it to be the same problem for the new plant that was rising Phoenix-like on the ashes of the old?

Kuron was acutely disillusioned. "I didn't used to be against atomic power plants," he said. "But now I can't help seeing how many mistakes are being made every day at Lagoona Beach. There's no pride of workmanship the way there used to be, and the quality control is for nothing. It wasn't a week after I got back to work, when they poured a floor of eighty yards of concrete and the whole thing gave way. It dropped three floors to the basement. If it hadn't been lunchtime, everyone working on it would have been killed. And what about these cracks in the floor of the main building? I'll bet you that they can't be corrected by high-pressure grouting, or anything else. The first Fermi plant was built well, and look what happened to it. Who knows what's gonna happen to this one. But I don't want to be around when they start up Fermi No. 2, especially with that concrete the way it is."

If the complaints had been confined to one Detroit Edison reactor, it would have been bad enough. But evidence was pouring in from all over the country about other lackluster and ineffectual construction. A secret 191-page AEC study was in the works that would soon reveal that among 30 operating reactors, 850 "abnormal occurrences" had been reported. No one knew how many were unreported. About four out of ten of these were traced to design and manufacturing errors. Others were the result of operator mistakes, faulty maintenance, poor control of the building process, executive goofs, and bad quality control. The AEC report said that these developments raised ". . . a serious question regarding the current review and inspection practices, both on the part of the nuclear industry and the AEC. This is particularly true when the increasing number of operating reactors which will be on-line in the 1980s and 1990s is considered."

This was the nightmare almost every thinking man in the nuclear fraternity was worrying about. When the year 2000 was reached, and with 1,000 reactors in operation, the nightmare would be even greater. A senior reactor safety engineer at the AEC's Idaho Falls installation made no bones about it when he said: "This is being advertised as a no-risk business, and that's not true. We don't know that reactors are unsafe, but we're concerned about their being as safe as the manufacturers would like you to believe. Maybe it's time the AEC told the public that if people want to turn their lights on, they are going to have to expect to lose a reactor now and then, and possibly suffer great dislocations and property losses as well."

But the AEC was not about to deal itself such a blow. Not only had it buried the figures of the WASH-740 update, but it had proceeded with the decision to pay out $3 million for a new probability study that might prove more palatable to the public. The acid test of this projected study, of course, would be whether or not the insurance companies would take over the coverage of the atomic power plants, and get the American taxpayer off the hook by eliminating the Price-Anderson insurance act.

The AEC itself, however, seemed to recognize that the study could, in fact, be useless. In stating the study's objectives, the AEC announcement declared: "It is recognized that the present state of knowledge probably will not permit a complete analysis of low-probability accidents in nuclear plants with the precision that would be desirable." But if this were so, critics began asking, why were $3 million being spent on it? More alarming was the AEC statement that the new study would "rely heavily on work currently being done by some reactor vendors . . . ," adding that industry's own Edison Electric Institute would be called on for information. It was quickly pointed out that this would be equivalent to instructing a high school senior to create his own final exam.

The man the AEC found to tackle the study was a professor at MIT named Norman Rasmussen. He was a strong supporter of fission atomic power plants, and made no bones about it. There was no question of his competence as a scientist, but he was put in a difficult position. It would be the insurance company underwriters who would be his ultimate judges and jury. If, for instance, the Rasmussen numbers showed that the chances of an atomic power plant accident were infinitely small, and the insurance companies did not immediately take over from the Price-Anderson government insurance, it would indicate clearly that the report amounted only to statistical gymnastics. As Ralph Nader was to say later: "If nuclear power is so safe, why won't the insurance industry insure it?"

Meanwhile, echoes of the Fermi No. 1 meltdown lingered. Remaining anonymous, an engineer at the Fermi project analyzed the accident: "Let's face it, we almost lost Detroit." His statement was circulated widely and it was hardly a reassuring thought. The fact remained, however, that they did not lose Detroit. Working with a reactor more complex than the SL-1 model at Idaho Falls, or the one at Windscale, or at Chalk River, McCarthy and his team were able to avoid what could have been an incredible disaster, by their planning, their expertise, their ingenuity, the low power level—and some luck.

It is often said that good ball players make their own luck. But why should the population of Detroit be faced with even the potential of such a disaster? Why should one of the best nuclear engineers in the world be faced with such a dangerous situation that he would say, with a measure of relief, that Detroit had been saved?

It was Eldon Alexanderson who would preside over the burial of Fermi No. 1. His job was to figure out how to take apart a core full of three and a half tons of radioactive uranium (speckled with enough plutonium to cause a decided uneasiness), thirty thousand gallons of radioactive sodium, and a vessel so bombarded with radiation that no one could enter it even in a protective suit. The bets were that the $4 million set aside to do the job wouldn't come anywhere near handling it. In Minnesota, after a decision was made to close it down, a far less complicated reactor had been given burial rites costing $7.5 million. One short-cut idea of Alexanderson's was to remove as much of the radioactive guts from the Fermi reactor vessel as possible, and seal it up. But it would still be hot. And there was no way to guarantee that corrosion wouldn't occur, or that water wouldn't leak out of the mausoleum in generations to come.

The nuances of the decommissioning were incredibly complex. Although the reactor was in the state known as subcritical, there could be a reactivity accident with little or no warning. Even a loud noise could actuate some mechanisms that could threaten the workers. The sodium could always hit air or water to create a tinder-box situation. It constantly had to be bathed in argon gas or nitrogen to avoid this. Health physicists continually had to monitor the reactor with Geiger counters. All kinds of contaminated equipment had to be dismantled and stored in what were called equipment decay tanks.

All unused penetrations into the reactor vessel had to be covered and seal-welded, including the heating and ventilating ducts. The uranium-packed subassemblies would have to be chopped up in three sections to ship out of the plant to a burial ground. But first they had to be submerged in "swimming pools" to cool off for months.

The casks used to ship the subassemblies were cylinders, nine feet in diameter, weighing eighteen tons each. They had to be sealed in a coolant, shielded with seven inches of lead, and mounted on flat-bed truck trailers to be shipped out. But since the casks were tested only for a thirty-foot fall and a thirty-minute fire, what might happen in the event of a shock impact or fire beyond those arbitrary limits was too frightening to contemplate. Dr. Marc Ross of the department of physics of the University of Michigan, has concluded that, if fire or impact distorted the shipping cask of a typical fuel shipment, the leakage of cesium from it would be particularly lethal, both directly through breathing it and indirectly through contamination of the food chain. Children, infants, and weakened adults would die if they were half a mile downwind from the accident.

Even the loading of the cut-up fuel assemblies was a precarious process. Each fuel unit was on the verge of becoming critical, even in the cooling water. But other problems had to be faced. Plutonium is so hazardous that no way has yet been found to permanently store it underground. It must be kept in recoverable containers and constantly monitored, while some solution is sought for the problem—a problem not just for the Fermi plant, but for the nationwide atomic power plant scene. There were six private burial grounds for contaminated materials in the United States. None of them would accept the irradiated sodium or the blanket material which had bred small amounts of plutonium. Grudgingly, the AEC agreed to consider handling the material itself, but no firm plan was made. The question dangled, and the price of dismantling soared above the $4 million mark.

The atmosphere inside the Fermi No. 1 plant became more and more like a mausoleum. "The decommissioning effort continues," Alexanderson said ruefully to a reporter in March of 1974. "Much of it is accompanied by a sinking feeling by staff personnel, as disassembly of many components gradually and irrevocably reduced the plant from the largest operating breeder reactor in the world to a fully decommissioned, partially dismantled status. It is very sad to see its demise. . . ."

To add to the sadness and desolation of the scene at Lagoona Beach, things were not going well with the construction of the new reactor, Fermi No. 2. Although its two new massive cooling towers, shaped like enormous smoke pots, loomed as a landmark seen from miles away, confidence in the project was slipping. Then, at the end of 1974, work on the second Fermi plant came to an abrupt halt. Financing for construction simply was not forthcoming. The whole complex, once filled with the clatter of trucks, cranes, and 1,600 helmeted workers, became a ghost town. Silence took over from the sound of pneumatic drills and power wrenches. The huge new reactor vessel, arriving by barge, had been eased into the reactor building and left there without its umbilical connections to the control room. The administration building stood as a red-iron skeleton. The concrete shells of the other buildings stood gaunt and empty, some with temporary sheet-iron roofs, amid the mud and the scattered disarray of a half-finished construction job. The scene caused one viewer to mumble: "There're more muskrats and rabbits around here than people. . . ."

On the other side of the site, the hollow shell of what had once been the Fermi No. 1 reactor sat by the gray waves of Lake Erie. In a shed next to the reactor building, triple decks of shiny black steel drums, all marked DANGER: RADIOACTIVE SODIUM, sat in a roped-off area—30,000 gallons of it that nobody wanted, or was willing to cart away. It was a problem no one had ever faced up to before, or really knew how to cope with.

Near the stacks of drums sat a box-like structure, about the size of an enlarged phone booth, marked cryptically by the work crew "Merlin's Box." It was here that the liquid sodium had been poured into the barrels through sealed pipes before it "froze" into the deadly chalky powder inside the drums. Some three hundred cubic feet of radioactive junk—cladding, rods, sawed-up metal from the spent guts of the reactor—were also scattered about the site, waiting to be carted away if takers approved by the AEC could be found. The radioactive blanket assemblies, specked with plutonium, still rested in "swimming pool" storage vaults. As welders made the final seals on the "hot" reactor vessel, plans were being made for the guards to set up their vigil—monitoring the shell with Geiger counters for generation after generation.

There were no trash barrels to hold this poisonous legacy. The dead Fermi breeder had spawned a $130 million ghost—a ghost that cannot be laid to rest.

AUTHOR'S EPILOGUE

Of all my experiences in preparing the lengthy research for this book there is one scene that stands out vividly. It occurred on a cold, wet January day. A strong wind was sweeping in from Lake Erie and I was going through the remains of the Fermi No. 1 plant with Eldon Alexanderson. He escorted me into the gaunt buildings that once had housed the huge breeder reactor in which so much hope had been placed for the peaceful use of the atom.

There was an errie hollowness to the buildings. The sound of our footsteps echoed loudly. Only a skeleton crew remained, and around the dome of the reactor vessel welders were sealing the last seams to close the empty core forever. A handful of engineers worked glumly in the control room, where panels from which instruments had been removed were left with gaping holes. In a connecting building, the deep "swimming pools" could be viewed from a narrow bridge that crossed them. Fuel assemblies from the reactor were hanging upright in the clouded water, carefully separated from each other by enough space so that no new fissionable reaction could start up.

Nearby, in a darkened storage area, were rows of fifty-gallon drums of radioactive sodium, six hundred of them piled three-deep in their shiny black casings, sitting mutely behind a rope barrier that warned against intrusion. This was the dangerous residue that nobody wanted—at any price. It symbolized the agonizing problem of how to dispose of the unwelcome wastes that were piling up at other reactors across the country. Being so close to them was not a comforting experience.

The research for the book led me on many crisscrossing paths: up into the cold but lovely country at Chalk River, Canada; out to the desolate mountains and flatlands in Idaho; over to the shores of the Irish Sea at Windscale; down to the lower Rhone in France; to Sweden and Switzerland. All through these journeys over many months, I listened to those who swore that nuclear energy would save the planet; I also listened to those who swore that it would destroy it. The more I traveled, the more I listened, and the more it became apparent that the answers being sought in this great debate would not be based on technical judgments. Instead, they would be judgments based on the indisputable facts that had emerged from two decades of experience with nuclear energy as a source of peacetime energy.

Any layman who cares to study these facts—and there are a jungle of them—can learn enough to make his own judgment. And he can do so without being told what to think by either the passionate supporters of nuclear energy or their equally passionate critics. Propaganda on both sides is heavy and loaded. But the facts can speak for themselves. They emerge clear and unassailable:

1. The AEC (now split into the NRC and ERDA) damage estimates regarding a major accident are conceptually catastrophic.
2. No one can buy insurance of any kind to cover such a catastrophe. No coverage at all is available for individuals, homes, or automobiles.

3. No solution has been found to handle the accumulation of poisonous radioactive wastes.
4. In one year, 1974–1975, nearly half of the more than fifty reactors in the United States had to be checked twice within a six-month period to see if there were cracks in their cooling pipes.
5. No realistic protection is available against terrorists seizure of nuclear plants or fuel, or against fuel-transport accidents.
6. Because of the billions of dollars allocated for nuclear fission power development, no *realistic* allocations have been available for developing alternate sources of energy such as solar, thermonuclear fusion, coal gasification and liquification, and others.

The idea for this book sprang from a suggestion by Bruce Lee, editor of the Reader's Digest Press, who mentioned to me at lunch one day that there ought to be a dramatic theme for a novel about a nuclear power plant that faces a meltdown crisis. The idea intrigued me and I worked up an outline. But the more we discussed the idea the more we worried that a fictional account might produce just a scare book. What we needed were facts. Thanks to the efforts of the Reader's Digest's Washington office, we obtained 5,000 pages of AEC documents that included the working papers of the Brookhaven-AEC meetings. Now we knew we were on the track of a significant story. The assignment was made.

My personal research began in Washington, when I talked to scientists, engineers, and executives of the AEC, and to members of the Joint Committee on Atomic Energy. I also talked to critics, who gave their side of the story. I visited nuclear plants in Maryland and Pennsylvania, and then went on to visit Canadian scientists in Ottawa and Chalk River, where the NRX and NRU accidents had taken place. I spent several weeks in Michigan, not only in Monroe and Lagoona Beach, but in Detroit, Ann Arbor, Lansing, and Midland, interviewing state and union officials, utility executives, state police, and health and legislative officials. At the enormous acreage of the Idaho Falls installations, where three men had died a gruesome death at the SL-1 reactor, I was taken around several experimental reactors. I traveled to Boston to get the point of view of both the pro and con forces at MIT. At each location, I picked up new leads for research.

As my travels continued, it became obvious that the story of the life and death of the Fermi breeder reactor, from its inception to its demise, was a more powerful one than any that could be invented. It also became apparent that the most damning evidence against the development of atomic power came, not from the critics, but from its most avid supporters. The minutes of the long-drawn-out Brookhaven-AEC meetings were appalling in their implications. What was even more appalling was the AEC's desire to keep them from reaching the public. Nearly every scientist and engineer I interviewed acknowledged there was no real solution to the burial or transportation of plutonium. And statement after statement by nuclear fission advocates showed that Murphy's Law—If anything can go wrong, it will—would occur sooner or later.

As I was working on the book, the $3 million Rasmussen study emerged. Suddenly, the public was being reassured. They were told that the chance of 1,000 people being killed by a reactor accident was about one in a million. This was the opposite of what my reporting had uncovered. There were other factors about the Rasmussen study that disturbed me. Sabotage was not even considered. Only two plants were used as pilots for the study, and they were light-water reactors. Breeder reactors—the most deadly of all—were ignored. Psychotic behavior and human error on the part of operators received no attention. All of the reservations of the WASH-740 Brookhaven report were bypassed. The dangers of fuel transportation, storage, and burial of radioactive wastes were skipped over. And while the report assumed complete evacuation in the area of damaged plants, no allowance was made for the futility of this operation.

After the Rasmussen report was issued, William Bryan, an aerospace engineer, pointed out in a congressional hearing that the study was an exercise in futility, because it had used analytical methods that had been completely discarded by the aerospace industry as unreliable. Ralph Nader described it in part as "fiction." Then an independent group of scientists headed by Rasmussen's fellow MIT colleague, Dr. Henry Kendall, prepared a review of Rasmussen's report. (Kendall and his staff were sponsored by the Union of Concerned Scientists and the Sierra Club.) Kendall's analysis, based on the methodology used in the Rasmussen report, indicated that a major nuclear power plant accident could kill or injure more than 120,000 people. In addition, the validity of the methodology itself was questioned. Taking the three reactors at Indian Point just outside of New York City for an example, the review concluded that no effective evacuation could possibly be made in the event of an accident where 16 million people lived within a forty-mile area.

I continued my research. The trail led to Windscale, England, where the accident with Windscale Pile No. 1 had occurred in 1957, sending radiation gauges soaring in London, three hundred miles away. But when I arrived, I found that only a short time earlier a second accident had occurred—a "blow back" in the plutonium-processing plant—and the entire installation had been temporarily closed.

Roaming the beautiful countryside of the Lake District, I was able to track down scientists, workers from the plant, and a number of union executives. I interviewed them in their homes, or in the pubs in Whitehaven. The seriousness of the second accident had come to public attention in an unexpected fashion. A housewife in Whitehaven was having a cup of tea in bed at 8 A.M. on October 25, 1973, waiting for her husband to come home from the night shift at the Windscale plant. Her husband arrived with strange news: Two health physicists were coming to their house to make a Geiger-counter reading of their bed and furniture. The wife barely had time to get dressed before they arrived. The results were negative, but the homes of other workers didn't fare as well.

Seven men, whose fate is not yet known, were placed under observation to check the dust from their lungs. (The smallest speck of plutonium can cause lung cancer.) Meanwhile, panic spread among the families of the others involved in the accident. John Noctor, the union representative at the plant, told me: "In the last two months, I have been approached by three women from Cleator Moor and Egremont who are no longer sleeping with their husbands."

Their fears had been confirmed when special bedsheets and pillowcases were issued to contaminated workers who had been told their sweat might contain radioactive poisons. One man reported for work only to have his underwear confiscated. Many of the thirty-three men reported to have plutonium poisoning were reluctant to talk about their family problems at a union meeting. Some were afraid to tell their wives for fear of losing them. But the wives learned about the problem from gossip at the corner shop. Today, the story is far from closed. Union officials are sponsoring long-term lawsuits for what they feel will inevitably be a series of death cases from plutonium poisoning.

All of this fortified a statement issued by Dr. Harold Urey, who had organized a group of leading scientists who declared that the handling and disposal of plutonium could never be solved, and that the billions earmarked for the development of the breeder reactor should be channeled into safer alternatives: the cleaning up of coal emissions and a crash development of solar energy. Such a transferral of funding away from nuclear development and into solar energy, they believed, would create a dramatic shift toward the practical realization of a power source that would combine safety with an inexhaustible, nonpolluting fuel supply.

But was this just a visionary dream? Many competent scientists did not think so, especially in the light of the faltering reactor program. As 1975 got underway, *The Wall Street Journal, Newsweek, The New York Times*, and other publications made many uncomfortable facts evident:

- The shutdown of roughly half the power reactors in the country for the second time in six months had made many investors jittery.
- Construction delays at nearly a hundred plants all over the country were clouding the industry's future.
- Costs of the planned demonstration breeder plant at Clinch River, Tennessee, had jumped from an estimated $700 million to more than $1.7 billion. In public discussions, government spokesmen do not acknowledge the existence of Fermi No. 1., but they indicate that breeder plants will be necessary in the future.
- The vulnerability of nuclear plants to earthquakes had not been solved.
- Emergency core cooling systems for light-water reactors still remained untested.
- Thousands of kilograms of plutonium are unaccounted for, some of which might be missing from government inventories.
- At least three new congressional hearings on nuclear safety were scheduled to be held in 1975.
- A referendum was being prepared in California to prohibit the building of nuclear power plants in that state.
- Belgium has instituted a moratorium on new nuclear facilities until its engineers produce a new reactor study. In France, four hundred scientists have asked the government not to allow any more nuclear plants until the public understands the hazards of such operations. The Swedish government is limiting nuclear operations in that country.

It was obvious that none of these negative factors could solve the critical need for increased energy production. If the nuclear power plants were not only dangerous but far more expensive than anyone believed, what could realistically replace them?

A consensus of opinions was emerging, even though there were many variations. Among them were:

1. There was growing agreement that the continuation of fission power plants might be suicidal, and that the program must be stopped, even if it meant a delay of fifteen or twenty years in easing the energy crisis.
2. To answer the short-term energy problem, the country's 500 to 800 years' supply of coal reserves should be used, accompanied by the swiftest possible program to convert coal to liquid and gas form so as to eliminate the sulfur dioxide emissions. Further, all strip mining should be tied into land reclamation.
3. There should be a voluntary cutback on energy use.
4. Funds going into fission power plants should be rechanneled into accelerated development of thermonuclear fusion.
5. There should be intensive research and development of alternate sources of energy, including geothermal, wind, and solar. Privately financed projects, such as an experimental house built by the Pennsylvania Power & Light Company show that a house can be both heated and cooled by solar energy for less than half the operating cost of conventional power sources.
6. There should be both a crash and a long-term program for the eventual total development of every form of solar energy, which would not only furnish an inexhaustible fuel source but also create less environmental damage.

What the consensus showed was that alternatives to fission power should be developed. But it would require a rethinking of priorities. Take, for example, the search for thermonuclear power, known as nuclear fusion (in contrast to fission). Unlike the fission power plants (such as the Fermi reactor and others planned), fusion produces little radiation waste to bury; it presents no danger of meltdown or explosion or breaking of the containment; and there is no problem of depletion of fuel, which the light-water reactors would be facing. There is enough deuterium—the basic fuel for fusion reactors—in the ocean waters to supply the potential demand for energy for more than a trillion years (longer than the estimated life span of the sun). According to AEC's Dr. Artlin Frass, tritium, the one villainous fission product of the thermonuclear fusion program, would create only 1/1,000,000 of the biological damage of the iodine-131 of the fission power plant.

What has held back the development of fusion power is that no final breakthrough has yet been made in harnessing this source of energy for peacetime use. One major reason for this has been a lack of research funds. By the mid-1960's, the entire budget for exploring fusion power was only around $20 million, about the cost of a National Football League franchise. The untested breeder reactor is slated to add up to a $5 billion cost to the taxpayer by 1985.

Those who have been working on thermonuclear power acknowledge that the highly complex problems of fusion, both theoretical and practical, are prodigious and far more intricate than with the light-water or breeder fission reactors. As yet, no one has made the breakthrough that would signal the mastery of the thermonuclear process. The main problem has been to reach a point where the energy produced by a fusion reactor is more than the energy that has to be put into it to create fusion. No one knows when that point will be reached. Many have been pessimistic that it can be achieved at all. The science of plasma physics is new and baffling. But one thing is certain: Vast sums of money are needed, and they have not been forthcoming from the AEC, or its successor agencies established in 1975.

But promising signs are beginning to appear. At Princeton, new thermonuclear experiments have jacked up the heating of plasma to more than double the previous levels. Oak Ridge, Los

Alamos, and the Lawrence Radiation Laboratory have reported progress in approaching the point where the energy put into the thermonuclear process might be matched by that created. KMS Fusion, in Ann Arbor, Michigan, reported in May, 1974, that its laser fusion process had actually reached the breakeven point. Projections from former AEC scientists show that an experimental thermonuclear reactor could be developed in the 1980's, with commercial fusion possible in the late 1990's. The significance of this projection—a possibility of commercially viable fusion by the late 1990's—must be weighed against the problems facing the breeder reactor program, whose commercial potential has been delayed from 1980 to 1990. In other words, if with only a minimum of federal funding the development of thermonuclear fusion is already a potential alternative, then it must be given more consideration than it has had in the past.

There are other alternatives. There is the energy of the sun itself. The world in the future will need some one quintillion British thermal units of energy per year. The sun, if it could be harnessed, could bring a gift of 3,600 times that amount to the surface of the earth. Neither the power from the sun nor thermonuclear fusion would rape the surface of the earth in a desperate search for fast-depleting fuel, nor would they burden it with enormous burial grounds of nuclear wastes.

But the problem with solar energy has been the same as that of thermonuclear power and other alternatives. Is it practical? Can it be realistically harnessed within the needed time span? William Heronemus, the engineering professor of the University of Massachusetts, is convinced that solar energy could become a reality in a very short time, providing national priorities were reordered. He blames the huge power group of the AEC, the utilities, and the reactor manufacturers for the dominance of fission power. He is convinced that our fission policy is a threat to the survival of mankind on earth, and he has showed figures to demonstrate that solar plants could be phased in to produce a significant amount of energy by the year 2000. But again, it depends on our priorities, which have, so far, been stacked against solar energy. Between 1950 and 1970, its development received a miserly $100,000 a year. And while, in 1975, the figure was raised to $50 million in research funds, it cannot compare to the $500 million that has been allocated for the breeder reactor program.

It took Frank Kuron and Tom Morgan nearly twenty years to comprehend the dangers nuclear energy presented to their community. Today, many more thousands across the country are beginning to realize the same thing. More than 100,000 signatures have been gathered for a petition against nuclear fission power plants by a Washington environmentalist group. The Federal Energy Administration could not find a single public-interest organization to testify in favor of nuclear power at hearings in Chicago. There is no evidence that ERDA (Energy Research and Development Administration) or the NRC (Nuclear Regulatory Commission), which took over the AEC functions in early 1975, will change the headlong rush to give priority to fission power.

There must be more public debate about the subject than there has been in years past. It is obvious that the government must create a more rational energy program than it has done to date. The tide of public opinion toward nuclear energy is shifting. If an enlightened public can now help a rechanneling of the billions of dollars spent for fission power toward the funding of research programs for alternative sources of power, there could be real hope for both easing the energy crisis and keeping our planet safe and clean.

ACKNOWLEDGMENTS

Grateful appreciation is expressed to the following people who helped supply material for this book by interviews and other research assistance:

Eldon Alexanderson, Hugh Anderson, Marion Anderson, Edward Bauser, Robert Bird, Robert Blackburn, Richard Black-ledge, Ruth Cashin, Walker Cisler, Patrick Clawson, David Comey, Bill Crozier, Irene Dickinson, Dan Ford, Ruth Fort, Raymond Fraley, Franklin Gage, Bjorn Gillberg, Bernard Giraud, Henry Gomberg, Leo Goodman, Virginia Gough, John Harkness, John Harris, Reginald Hayden, Robert Hirsch, Carl Hocevar, Bob Hughes, Frank Ingram, Wayne Jens, Henry Kendall, Chi-roro Kikuchi, Doug Kirkpatrick, Herbert Kouts, Frank Kuron, Sheila Lalwani, Virginia Lawton, Howard Lipton, Amory Lovins, James Lyman, Joseph Marrone, William Maxwell, Walter McCarthy, Helia Miido, Tom Morgan, Ralph Nader, Thomas Nemzek, John Noctor, Eagan O'Connor, Gil Omenn, Walter Patterson, William Ralls, Norman Rasmussen, Dixie Lee Ray, Judith Robinson, Marc Ross, Richard Sandler, Bridgid Scanlon, Vern Schneider, Milton Shaw, Mary Sinclair, Earl Sliper, Conrad

Spohnholtz, Vern Strickland, Hal Tracey, Marc Valantin, Joseph Van Hart, William Voight, George Weil—and numerous others whose help is much appreciated.

I would also like to thank my editors at Reader's Digest Press, Bruce Lee and Nancy Kelly, for their editorial guidance.

BIBLIOGRAPHY

Advisory Committee on Reactor Safeguards. June, July, 1955. Minutes of 13th meeting.

Alabama Journal. August, 1972. "The Economics of Atomic Power."

Anderson, M. 1974. *Fallout on the Freeway: The Hazards of Transporting Radioactive Wastes in Michigan.* PIRGIM.

Atomic Energy of Canada, Ltd. 1973. *Heavy Water: A Layman's Guide.*

Atomic Energy Office. 1957. *Accident at Windscale No. 1 Pile on 10th October 1957.* Presented to Parliament by the Prime Minister. Her Majesty's Stationery Office.

Atomic Industrial Forum, Inc. *What's a Breeder?*

———. 1973. *Nuclear Power Expected to Save U.S. Equivalent of 3 Billion Gallons of Oil This Winter.*

———. *Nuclear Industry.* 1973. Vol. 20. No. 3.

———. *Nuclear Industry.* 1973. Vol. 20, No. 6.

———. *Nuclear Industry.* 1973. Vol. 20, No. 7.

Atomic Power Development Associates, Inc., Power Reactor

Development Company. February 1967. Report on October 5, 1966, Fuel Damage Incident at the Enrico Fermi Atomic Power Plant.

Asimov, I. 1972. *Electricity and Man.* U.S. Atomic Energy Commission.

———. 1972. *Worlds Within Worlds: The Story of Nuclear Energy.* Vol. 1. U.S. Atomic Energy Commission.

———. 1972. *Worlds Within Worlds: The Story of Nuclear Energy.* Vol. 2. U.S. Atomic Energy Commission.

Bernstein, J. 1969. *The Elusive Neutrino.* U.S. Atomic Energy Commission.

Boutrais, M., et al. 1973. *Sodium Analysis and Purification. Aerosols Trapping in Phénix Reactor.* Phénix Project. Pont-St-Esprit, France.

British Nuclear Energy Society. March 1974. *International Conference: Fast Reactor Power Stations.*

———. 1974. *The Status of the Fuel Elements for SNR 300 with Consideration of the Technical and Economical Aspects.* Paper to be presented at the British Nuclear Energy Society International Conference on Fast Reactor Power Stations.

———. Conference on Fast Reactor Power Stations, Institution of Civil Engineers. 1974. *United States LMFBR Demonstration Plant Activities.*

Burnham, D. January, 1975. "Atomic Reactors May Have to Shut." *New York Times.*

Business Week. October, 1973. "Building a Fire under Magnetohydrodynamics."

CBS News "Face The Nation." December, 1973. Guest: Ray, D. L., chairman, Atomic Energy Commission.

Chernick, C. 1967. *The Chemistry of The Noble Gases.* U.S. Atomic Energy Commission.

Cisler, W. October, 1970. Testimony before Michigan Public Service Commission

Cockermouth Urban District Council, The. 1970. *Cockermouth: Birthplace of William and Dorothy Wordsworth.* Cumberland Newspapers, Ltd.

Columbia Record, The. November, 1972. "Energy Facility Closes."

Commissariat à l'Energie Atomique et Electricité de France. 1971. *Prototype Fast Reactor Power Station Phénix.*

Courtenay, A. 1973. *Let's Halt Awhile in Great Britain.* Hastings House, Publishers, Inc., New York.

Cowper, G. *Health Physics Instrumentation: Atoms for Development.* Atomic Energy of Canada, Ltd.

Curtis, R., and Hogan, E. 1969. *Perils of the Peaceful Atom.* Ballantine Books, Inc., New York.

Daniels, F. 1964. *Direct Use of the Sun's Energy.* Ballantine Books, Inc., New York.

Detroit Edison Company. 1963. *Chronological Record of Atomic Power Development Activities.*

Detroit Evening News. February, 1963. "Big Names Build Atomic Project."

———. November, 1969. "Visited by Soviets."

Detroit Free Press. January, 1957. "Stage Set for A-Plant Battle."

———. January, 1957. "Air Base May Figure In A-Feud."

———. January, 1963. "The Problems Were Many at A-Plant."

———. September, 1965. "Expert O.K.'s Monroe A-Plant."

———. August, 1966. "Fermi Atom Plant Yields Electricity in First Trial Run."

———. November, 1967. "Is a Beer Can to Blame for Fermi Plant Shutdown?"

———. November, 1967. "Is Beer Can Behind A-Plant Failure?"

———. August, 1972. "AEC Decides to Shut Fermi Nuclear Plant."

———. December, 1972. "Atom Plant Will Close—But How?"

———. February, 1974. "A-Plant Problems Mushroom: Utility's Nuclear Nightmare."

Detroit News. November, 1966. "Could Fermi Plant Level Detroit?"

———. February, 1970. "Fermi Atom Plant to Resume Work."

———. October, 1970. "Baby Death Rise Linked to A-Plant."

———. April, 1974. "Annual Meeting Hits a Sour Note. Edison Holders List Complaints."

Donnell, A. "Key to Nuclear Fuel Conservation . . . The Enrico Fermi Power Plant." *Allis-Chalmers Electrical Review.*

Donnelly, W. 1972. *Atomic Energy: Increase in the Price-Anderson Indemnification.* The Library of Congress.

Dunbar, W. July, 1964. Nuclear Excursion Accidents: Their Causes and Consequences. Unpublished paper.

Electric World. January, 1973. "Fermi I: Success or Failure?"

Enrico Fermi Atomic Power Plant, Power Reactor Development Company. April, 1962. License application. Technical Information and Hazard Summary Report.

———. August, 1965. Radiation Emergency Procedures Operating Procedure No. 704.

Evening Mail. (Whitehaven, England.) October 23, 1973. "Some Wives Fear For Their Marriages—Windscale Union Convenor."

———. October 25, 1973. "Whitehaven Wife in Breakfast Scare as Atom Men Test Bed."

Federal Disaster Assistance Program Handbook. July, 1972. Executive Office of the President, Office of Emergency Preparedness.

Feldes, J. G. November, 1962. Testimony to U.S.A.E.C. for Power Reactor Company on application for Provisional Operating License.

Frigerio, N. 1966. *Your Body and Radiation.* U.S. Atomic Energy Commission.

Gifford, F. A. Jr. Advisory Committee on Reactor Safeguards. October, 1962. Letter to Seaborg, G. T., chairman, U.S. Atomic Energy Commission.

Gillberg, B. November, 1973. "Sweden: Doubts of Nuclear Power." *Environment.*

Glasstone, S. 1968. *Controlled Nuclear Fusion.* U.S. Atomic Energy Commission.

———, and Thomas, S. 1970. *Atomic Energy and Your World.* U.S. Atomic Energy Commission.

———. 1972. *Inner Space: The Structure of the Atom.* U.S. Atomic Energy Commission.

Gofman, J. 1972. *The Case for a Nuclear Moratorium.* Environmental Action Foundation.

Gomberg, H. J., et al. July, 1957. Report on the Possible Effects on the Surrounding Population of an Assumed Release of Fission Products into the Atmosphere from a 300-Megawatt Nuclear Reactor Located at Lagoona Beach, Michigan. Atomic Power Development Associates, Inc.

———, and Charatis, G. April, 1974. Recent Results of Laser Induced Fusion Tests at KMS Fusion. Paper presented at ANS First Topical Meeting on The Technology of Controlled Nuclear Fusion, San Diego, Calif.

Goodman, L. September, 1965. Report on Accidents and Malfunctions Changes and Modifications of Detroit Edison Fast Reactor.

Gould, W. 1973. *Energy Options for the Future.* Atomic Industrial Forum, Inc.

Green, J. July, 1974. "A New Era in Energy?" *Detroit Sunday News Magazine.*

Greenwood, J. 1964. Contamination of the NRU Reactor in May, 1958. Atomic Energy of Canada, Ltd.

Gschneidner, K. 1964. *Rare Earths: The Fraternal Fifteen.* U.S. Atomic Energy Commission.

Hammond, A. L. November, 1971. "Breeder Reactors: Power for the Future." *Science.*

———. 1972. "The Fast Breeder Reactor: Signs of a Critical Reaction." *Science.*

Hanauer, S., chairman, Advisory Committee on Reactor Safeguards. December, 1969. Letter to Seaborg, G. T., chairman, U.S. Atomic Energy Commission.

Hanford Engineering Development Laboratory. 1972. Fast Flux Test Facility Design Safety Assessment.

Hansen, G. 1958. Elementary Fast Neutron Critical Assemblies: Proceedings of the Second U.N. International Conference on Peaceful Uses of Atomic Energy. Vol. 12.

Haubold, W., and Jung, J. 1974. Impurity Monitoring Techniques for LMFBR Sodium Systems. Paper presented at British Nuclear Energy Society International Conference on Fast Reactor Power Stations.

Hicks, E., and Menzie, D. 1965. Theoretical Studies on the Fast Reactor Maximum Accident. Proceedings of the Conference on Safety, Fuels, and Core Design in Large Fast Power Reactors. Argonne National Laboratories.

Hogerton, J. 1963. *Nuclear Reactor.* U.S. Atomic Energy Commission.

———. 1963. *Atomic Fuel.* U.S. Atomic Energy Commission.

Hughes, R. February, 1973. "Fermi A-Plant Gingerly Dismantled." United Press International.

Hurst, D. 1964. The Accident to the NRX Reactor. Part II. Atomic Energy of Canada, Ltd.

Jackson, B. 1971. *Epicenter.* Berkley Publishing Corporation, New York.

Joint Committee on Atomic Energy. April, 1959. Indemnity and Reactor Safety: Hearings before the J.C.A.E. U.S. Government Printing Office.

———. 1961. SL-1 Accident: AEC Investigation Board Report. U.S. Government Printing Office.

———. April, 1962. Indemnity and Reactor Safety: Hearings before the Subcommittee on Research, Development, and Radiation of the J.C.A.E. U.S. Government Printing Office.

———. 1971. Controlled Thermonuclear Research. Hearings

before the J.C.A.E. Part I. U.S. Government Printing Office.

———. February, 1972. AEC Authorizing Legislation Fiscal Year 1973: Hearings before the J.C.A.E. U.S. Government Printing Office.

———. September, 1972. Liquid Metal Fast Breeder Reactor Demonstration Plant: Hearings before the J.C.A.E. U.S. Government Printing Office.

———. 1973. Understanding the "National Energy Dilemma." U.S. Government Printing Office.

———. April, 1973. Membership, publications, and other pertinent information through 92d Congress, 2d Session. U.S. Government Printing Office.

———. 1973. Current Membership. U.S. Government Printing Office.

———. 1973. AEC Authorizing Legislation Fiscal Year 1974. Part III. Hearings before the J.C.A.E. U.S. Government Printing Office.

Kastner, J. 1968. *The Natural Radiation Environment.* U.S. Atomic Energy Commission.

———. 1973. *Nature's Invisible Rays.* U.S. Atomic Energy Commission.

Klickman, A. E. Atomic Power Development Associates, Inc. January, 1971. Letter to Sandler, R., Senate Commerce Committee.

KMS Industries. May, 1974. "KMS Announces Major New Progress in Obtaining Energy from Laser-Fusion."

LeCompte, R., and Wood, B. 1968. *Atoms at the Science Fair.* U.S. Atomic Energy Commission.

Lewis, R. 1972. *The Nuclear-Power Rebellion.* The Viking Press, Inc., New York.

Lewis, W. 1963. *The Accident to the NRX Reactor on December 12, 1952.* Atomic Energy of Canada, Ltd.

LMFBR Nuclear Safety Program Annual Report, 1971. U.S. Atomic Energy Commission.

Loewenstein, W., and Okrent, D. 1958. Physics of Fast Power Reactors: Proceedings of the Second U.N. International Conference on Peaceful Uses of Atomic Energy. Vol. 12.

Love, S. 1972. The Case for Reassessing the Fast Breeder. Testimony. Environmental Action Foundation.

Lovins, A. November, 1973. *World Energy Strategies: Facts, Issues, and Options.* Earth Resources Research, Ltd.

———. March, 1973. "The Case Against the Fast Breeder Reactor." *Bulletin of the Atomic Scientists.*

Lyerly, R., and Mitchell, W. 1967. *Nuclear Power Plants.* U.S. Atomic Energy Commission.

Manchester Guardian. October 12, 1957. "Uranium Becomes Red Hot in Atomic Pile: Workers warned after Windscale accident but 'no hazard to public.' "

———. October 13, 1957. "Ban on Farm Milk Near Windscale: Radio iodine content found to be six times too much. Threat to children?"

———. October 16, 1957. "Milk from 200 Square Miles Goes into the Sea."

———. October 17, 1957. "Workers at Calder Hall Ask for Information: New cattle-slaughtering precautions."

———. November 9, 1957. "Why Windscale Went Wrong: Errors of Men and Instruments."

Marrone, J. 1967. "Nuclear Insurance—Future of the Nuclear Liability Pools." *The National Underwriter.*

———. 1971. "Nuclear Liability Insurance—A Brief History Reflecting the Success of Nuclear Safety." *Nuclear Safety.*

McBride, J. 1971. *The End of the Cycle: Transportation, Reprocessing, and Waste Management.* Atomic Industrial Forum Workshop.

McCarthy, W. J., Jr. November, 1962. Testimony to U.S. Atomic Energy Commission for Power Reactor Company on application for Provisional Operating License.

McCarthy, W. J., Jr., and Jens, W. H. November, 1967. "Enrico Fermi Fast Breeder Reactor: The reactor's fuel damage incident and its significance to future design." *Nuclear News.*

McCullough, C. R., chairman, Advisory Committee on Reactor Safeguards. July, 1955. Letter to Fields, K. E., U.S. Atomic Energy Commission.

———. June, 1956. Letter to Fields, K. E., U.S. Atomic Energy Commission.

McIlhenny, L. 1968. *Careers in Atomic Energy.* U.S. Atomic Energy Commission.

McNamara, P. August, 1957. Remarks made in the U.S. Senate.

Meyer, R., et al. 1965. A Parameter Study of Large Fast Reactor Meltdown Accidents. Proceedings of the Conference on Safety, Fuels, and Core Design in Large Fast Power Reactors. Argonne National Laboratories.

Michigan (State of) Water Resources Commission. December, 1962. Statement of position re operation of breeder reactor by Power Reactor Development Company.

Michigan State Health Commissioner. December, 1962. Statement of position with reference to the application of the Power Reactor Development Company for a provisional operating license for the nuclear facility at Lagoona Beach.

Michigan, Township Board, Frenchtown Township, Monroe County. December, 1962. Statement of Township Board's position on the application of the Power Reactor Development Company for provisional operating license.

———. October, 1973. State of Michigan Nuclear Power Plant Emergency Plan. (Working Draft.)

Michigan Weekly Free Press. January, 1974. "Nixon: AEC to Set Radiation Standards."

Miller, J. N. June, 1972. "Just How Safe Is a Nuclear Power Plant?" *Reader's Digest.*

Mitchell, W. 1972. *Grasmere and the Wordsworths.* Galava Printing Company, Ltd., England.

Morgan, T. April, 1972. Letter to Miner, S. U.S. Atomic Energy Commission.

Mozola, A. J. 1970. Geology for Environmental Planning in Monroe County, Michigan. State of Michigan Geological Survey Division.

Nader, R. 1973. Letter to the Hon. Dixy Lee Ray, chairman, U.S. Atomic Energy Commission.

Nicholson, R. Methods for Determining the Energy Release in Hypothetical Reactor Meltdown Accidents, APDA-150.

Nolan, J., et al. 1974. Fast Flux Test Facility: Paper for presentation at British Nuclear Energy Society International Conference on Fast Reactor Power Stations, London, 1974.

Novick, S. June, 1966. "Reactions to Nuclear Reactors." *Science.*

———. June, 1967. Bad Day at the Fermi Plant: A Little-Reported Atomic Accident. *Scientist and Citizen.*

———. 1969. *The Careless Atom.* Houghton Mifflin Company, Boston, Mass.

———. July/August, 1974. "Nuclear Breeders." *Environment.*

Nuclear Energy Liability Insurance Association and Mutual Atomic Energy Liability Underwriters. 1973. *Safety in Nuclear Industry Reflected by Insurance Refunds.* Insurance Information Institute.

Nuclear Energy Liability Policy. Nuclear Energy Liability Insurance Association.

Okrent, D., et al. 1955. Fast Reactor Physics: Proceedings of the U.N. International Conference on Peaceful Uses of Atomic Energy. Vol. 5.

Pennsylvania Insurance Department. 1973. Citizen's Bill of Rights and Consumer's Guide to Nuclear Power.

Perry, H. March, 1974. "The Gasification of Coal." *Scientific American.*

Pesonen, D. E. October, 1965. "Atomic Insurance. The Ticklish Statistics." *The Nation.*

Plymouth County Nuclear Information Center. "Do You Know What Plutonium Is?"

Pollard, W. *The Mystery of Matter.* U.S. Atomic Energy Commission.

Power Reactor Development Company. November, 1966. Letter to Price, H., director of Regulation, Atomic Energy Commission.

——— October, November, December, 1966. Monthly reports on operation of Enrico Fermi Atomic Power Plant.

———. Enrico Fermi Atomic Power Plant. December, 1968. Report on the Fuel Melting Incident in the Enrico Fermi Atomic Power Plant on October 5, 1966. Submitted to Atomic Energy Commission.

———. Docket No. 50-16. Application to U.S. Atomic Energy Commission for amendment of Provisional Operating License.

———. 1972. Request to Atomic Energy Commission for extension of expiration date of Provisional Operating License.

———. May, 1973. Request for Extension of License Expiration Date.

Price, E. R., Director, Division of State and Licensee Relations. Amendments to Indemnity Agreement No. B-20. Amendment No. 2, July 1962; No. 3, July 1963; No. 4, June 1964; No. 7, January 1966; No. 9, July 1966.

Radiation Research Society. May, 1971. Papers from 19th Annual Meeting.

Ray, J. H., et al. July, 1957. Calculated Criticality of the Enrico Fermi Reactor during a Hypothetical Slow Meltdown Accident. Nuclear Development Corporation of America.

Rocks, L., and Runyon, R. 1972. *The Energy Crisis.* Crown Publishers, Inc., New York.

Rohrer, E., et al. 1961. *New Criticality Excursion Machines.* U.S. Atomic Energy Commission.

Ross, M. 1974. The Possibility of Release of Cesium in a Spent-Fuel Transportation Accident. Unpublished paper.

Singleton, A., Jr. 1968. *Sources of Nuclear Fuel.* U.S. Atomic Energy Commission.

Sternglass, E. 1972. *Low-Level Radiation.* Ballantine Books, Inc., New York.

Stockholm Conference Eco. Various issues. 1972.

Supreme Court of the United States. October term, 1960. Power Reactor Development Company, Petitioner v. International Union of Electrical, Radio and Machine Workers, AFL-CIO, et al.

Thalgott, F. 1957. *Fast Reactor Safety: EBR-1.* U.S. Atomic Energy Commission.

———, et al. 1958. *Stability Studies of the EBR-I, Marks i to iii.* International Atomic Energy Agency.

Thompson, T., and Beckerley, J. (eds.) 1964. *The Technology of Nuclear Reactor Safety.* Vol. 1. The M.I.T. Press, Cambridge, Mass.

Tinker, J. March, 1973. "Breeders: risks man dare not run." *New Scientists.*

University of Chicago (The). 1973. Energy for the Future: A Symposium.

U.S. Government House of Representatives Report N1883. 1965. Extending and Amending the Price-Anderson Indemnity Provisions of the Atomic Energy Act of 1954, As Amended.

U.S. Atomic Energy Commission. September, 1961. Amendment to Indemnity Agreement No. B-20 between Power Reactor Development Company and the Atomic Energy Commission.

———. Advisory Committiee on Reactor Safeguards. October, 1962. Letter to Seaborg, G. T., Atomic Energy Commission.

———. The New Force of Atomic Energy—Its Development and Use.

———. March, 1962. Indemnity Agreement No. B-20 between

Power Reactor Development Company and the Atomic Energy Commission.

U.S. Atomic Energy Commission. December, 1962. Order Denying Motion for Postponement of hearing in the matter of Power Reactor Development Company's Application for Provisional Operating License.

———. 1968. Nuclear Power Reactor Plant Siting.

———. March, 1969. Proceedings at Atomic Energy Commission Symposium, Lawrence Radiation Laboratory. Biological Implications of the Nuclear Age. Vol. 16.

———. 1970. Atomic Energy Basics.

———. June, 1970. Request for Authorization of Change No. 23 in the Technical Specifications in the matter of Power Reactor Development Company Provisional Operating License No. DPR-9.

———. August, 1971. Supplement to Safety Evaluation by the Division of Reactor Licensing in the matter of Enrico Fermi Atomic Power Plant Unit 2.

———. 1971. Science information available from the Atomic Energy Commission.

———. 1972. WASH-1139. Nuclear Power 1973-2000.

———. 1972. WASH-1509.

———. March, 1972. Draft Environmental Statement, Enrico Fermi Atomic Power Plant Unit 2.

———. April, 1972. Environmental Statement: Liquid Metal Fast Breeder Reactor Demonstration Plant.

———. August, 1972. Power Reactor Development Company. Docket No. 50-16 Denial of Application for License Extension and Order Suspending Operations.

———. 1973 Financial Report.

———. Reports File EF2—1,796. May, 1973. Enrico Fermi Atomic Power Plant Unit 2, Detroit Edision Company. Dewatering of RHR Complex During Construction.

———. July, 1973. WASH-1250 Report: The Safety of Nuclear

Power Reactors (Light Water-Cooled) and Related Facilities. (Final Draft.)

U.S. Atomic Energy Commission. 1973. National Reactor Testing Station.

———. 1973. Atomic Energy Commission Research and Development Laboratories.

———. Nuclear Terms: a Glossary.

Vollrath, J. A. February, 1972. *Memories of the Monroe Piers.* The Monroe County Library System.

Webb, R. E. January, 1971. Letter to Klickman, A. E., Atomic Power Development Associates, Inc.

———. January, 1971. Some Autocatalytic Effects During Explosive Power Excursions in Fast Nuclear Reactors.

———. July, 1973. The Explosion Hazard of the LMFBR.

Weil, G. L. July, 1971. Nuclear Energy: Promises, Promises.

———. July 1973. Letters to Augustine, R. E., National Intervenors.

Weinberg, A. 1972. *The Safety of Nuclear Power.* Atomic Industrial Forum, Inc.

INDEX

ABOUT THE AUTHOR

John G. Fuller is the author of the best sellers *Incident at Exeter, The Interrupted Journey*, and *The Day of St. Anthony's Fire*, as well as the highly acclaimed *200,000,000 Guinea Pigs* and *Arigo: Surgeon of the Rusty Knife*. His latest book, *Fever!*, won honorable mention by The New York Academy of Sciences. Fuller is also a playwright and has written, directed, and produced a number of TV documentaries. He lives in Weston, Connecticut.